# BALLISTIC

# BALLISTIC

The New Science of Injury-
Free Athletic Performance

## HENRY ABBOTT

Illustrations by John Early

**W. W. NORTON & COMPANY**

*Independent Publishers Since 1923*

For information about permission to reproduce selections from this book, write to
Permissions, W. W. Norton & Company, Inc., 500 Fifth Avenue, New York, NY 10110

For information about special discounts for bulk purchases, please contact
W. W. Norton Special Sales at specialsales@wwnorton.com or 800-233-4830

Manufacturing by Lakeside Book Company
Book design by Beth Steidle
Production manager: Julia Druskin

ISBN 978-1-324-05013-1

W. W. Norton & Company, Inc., 500 Fifth Avenue, New York, NY 10110
www.wwnorton.com

W. W. Norton & Company Ltd., 15 Carlisle Street, London W1D 3BS

10 9 8 7 6 5 4 3 2 1

*For Jessica,*
*Molly,*
*Duncan,*
*and glutes.*

The Church says: The body is a sin.
Science says: The body is a machine.
Advertising says: The body is a business.
The Body says: I am a fiesta.

—Eduardo Galeano

# Contents

# BALLISTIC

# INTRODUCTION

LEGENDARY NBA COACH Phil Jackson won his tenth NBA title at about 10:30 p.m. on June 14, 2009. A couple of hours later, his blue suit jacket still a little champagne-sticky, the Lakers' sixty-three-year-old leader exited Orlando's Amway Arena in a phalanx of his adult children.

I happened to be a few steps behind. What I remember most aren't the hats with the Roman numeral ten—X—his kids wore, nor their joyous expressions. Instead, it's the pained look on Phil's face. The greatest coach in the history of the game waddled the endless arena hallway with surgically fused vertebrae, balloon stents in his arteries, and none of his original hips or knees. The family shuffled at Phil's pace. After playing for twelve NBA seasons and coaching for twenty-three, Phil could barely walk.

He's hardly alone. According to the Centers for Disease Control and Prevention, even before the pandemic 71 percent of young Americans "would not be able to join the military if they wanted to," because they're not fit enough. Twelve percent of Americans have serious difficulty walking or climbing stairs. About 31 million Americans age fifty or older "get no physical activity beyond that of daily living." In 1969, half of American kids walked or biked to grade school; in 2009, it was less than 13 percent.

In short, we can't move—and it's killing us. The American College of Sports Medicine submitted a report to the US Secretary of Health and Human Services called *Exercise Is Medicine*, which notes that physical activity improves bone health, cognitive function, heart disease, stroke, many cancers, dementia, sleep, anxiety, depression, multiple sclerosis, hypertension, quality of life, and mortality. The World Health Organization says inactivity causes five million deaths a year.

Talk to your doctor, they say, before starting an exercise program. With immobility the fourth-leading cause of global death, who do you talk to before *not* exercising? Joining a gym or going for a run seems, by comparison, safe and healthy. But, we all understand, along the way, something might hurt. From 2008 to 2018, the rate of Americans seeking medical help for injuries (at least, those not caused by car accidents) climbed in almost every age group, gender, and subcategory.

We can be bloody-minded about how we move, pushing our bodies like drill sergeants. Trainers describe a trend: People tend to do what they did in high school. Put on thirty extra pounds in the pandemic? You might dig out those decades-old soccer cleats, join a league, and expect to race through the midfield like the old days. NBA players do the same thing: Hip starts hurting three years into your professional career? The first thought for many is to bring back the weight-lifting routine they had when they were seventeen, when they hammer dunked on everybody and felt amazing.

But that high-schooler had a certain lung capacity, hip range of motion, and glute strength, not to mention the spine of someone who hadn't yet spent decades working on a laptop. Dead-sprinting the left wing of a soccer field, or cutting through the lane to catch a lob on the basketball court, wasn't much to ask of a teenaged body with bounce, wind, and muscle to spare.

It's common sense that if you park a car for twenty years, you might need to take it to a mechanic before hitting the open road. The same is true of our bodies. But we're more complicated, needing more than an oil change and four new tires. How do you maintain an athletic body?

Reliable answers to that question can be so maddeningly elusive that they seem not to exist at all. Injuries feel inevitable to weekend joggers and professional athletes alike. Bad backs, trick knees, degraded hips—who doesn't have at least one body concern and a grab bag of theories about stretching, yoga, massage, physical therapy, cupping, weight lifting, or whatever else feels right? But when it comes to evidence-based sports injury prevention, there's just not a lot to go on. If you want a solution to move and feel better, you probably feel like I did for most of my life: a little lost.

In my decades as an athlete, I have suffered plenty of injuries, and I hated every one of them—because I love to move. In high school I ran track and cross-country, raced on skis, and played a little soccer. In the decades since, I've biked across states, run all kinds of marathons, and taken a thousand or so hard workout classes. And along the way, I've had wildly frustrating experiences, doctor-hopping, looking for relief from hip muscles in revolt, Achilles tendinitis, and a whopper of a lumbar crisis.

That pain felt like a personal failing, until I noticed identical things happening to the best athletes in the world. For the last quarter century, I've worked as a journalist covering the free-moving sport of basketball. The NBA boasts some of the world's most extraordinary, highest-paid athletes. But even as data made the game more exciting and sent viewership soaring, the athletes worth hundreds of millions spent more and more time on the bench.

I'll never forget the day that a physical therapist recommended that, before sitting down on a chair, in deference to my hips, I first put down a pad shaped like a wedge of cheese. I hesitated. Was this really how I'd roll into restaurants for the rest of my life? The next day I spotted something hilarious during a Laker game: As LeBron James checked out of the game, a trainer slipped an identical cheese wedge onto his seat.

The NBA devours athletic bodies. I grew up in Oregon, where Bill Walton won the 1977 NBA championship—then sued the Portland Trail Blazers for mismanaging his injuries. He missed three of the next four seasons. Larry Bird missed almost two hundred games, then retired prematurely. Citing the toll on his body, Michael Jordan sat out almost four seasons because he was hurt, playing baseball, or retired. Kobe Bryant, Dwyane Wade, Shaquille O'Neal, and Stephen Curry each sat out hundreds of games

In the mid-1990s, Grant Hill got more All-Star fan votes than Jordan, then suffered a brutal string of injuries that he attributes to "archaic" training methods. Ankle surgery blossomed into a staph infection that nearly killed him; Hill was never more than a role player after that. Over the course of his career, Hill missed more than six seasons' worth of regular-season games. (In a league where a typical career lasts three years, Hill sat in street clothes for two NBA careers.)

Once upon a time, baseball's Cal Ripken Jr. played 2,632 consecutive games and lulled sports fans into believing that a human can be athletically elite and stable over the long term. Since then, players have become measurably bigger, stronger, and faster. But these increases seemed to reduce their longevity, mental health, and robustness. For all we have learned about cutting-edge performance, we seem to have learned very little about preventing injury. Sports are still lousy with meatheaded "rub some dirt on it" thinking that seamlessly transitions into "next man up" when it all goes awry.

Like race cars, NBA players are in the garage often and sometimes badly wrecked. One expert casually told me that the NBA has a 100 percent injury rate. "Every game you don't get hurt," he said, "is a game closer to your next injury." Many of the league's would-be superstars never reach their would-be primes: Anfernee Hardaway drew comparisons to Magic Johnson, but knee injuries meant he played in his last All-Star Game at age twenty-six. Other than a five-game comeback in 2010, Yao Ming stopped playing at age twenty-eight. We may never know what Zion Williamson could have been. A dozen big-name players, including Giannis Antetokounmpo, Jimmy Butler, and Kawhi Leonard missed games in the 2024 playoffs.

Magnetic resonance imaging and surgical techniques march forward; surgeries to repair a torn Achilles or anterior cruciate ligament have improved. Doctors have incredible new insights into what happens after an injury has occurred, but that doesn't lead to fewer injuries in the first place. The crisis continues.

At the 2014 MIT Sloan Sports Analytics Conference, I was a last-minute invitee to a private-room dinner hosted by former Goldman Sachs executive Dave Heller, then one of the investors who ran the Philadelphia 76ers. A few dozen people ate around a ring of tables as Heller steered the conversation like Oprah. The conference is about how big data can improve sports decision-making. By 2014, wearable devices, sleep studies, and player tracking data had started a conversation that we continued, around Heller's table, about player health.

I said something about making sure the league doesn't "break Derrick Rose." Rose was the Chicago Bulls' next great guard after Jordan. Rose wasn't a great shooter, nor was he especially tall, but he had won an MVP award doing things that made fans (and even sportswriters) love basketball, heroically taking on bigger defenders in the paint and launching himself to score in spectacular ways.

The Bulls star was, at the time, rehabbing a torn left ACL. Somewhere off to my right, a white man in a suit piped up. He had the slick smugness of Kieran Culkin in *Succession*. I never learned his name, but I'll never forget that he said, "What do I care if I break Derrick Rose? I'll get another one."

It was outrageous, racist, and, upon reflection, not far from everyday NBA thinking. The Bulls *did* break Derrick Rose, and they *did* get another one. High-scoring guard Zach LaVine, measured as one of the finest athletes in league history, tore his ACL, too.

To better understand why all these players were getting hurt, I began a side project researching the science of injuries, reading studies, interviewing doctors, and attending conferences. Meanwhile, the problem had grown personal. At Heller's dinner, I had been sitting awkwardly to avoid shooting pain. A few months earlier, I had tweaked something while running a half-marathon. Now, stepping off curbs, kicking balls, sleeping, sprinting, coughing, and sneezing—almost every ordinary, everyday action infused my body with thunderous jolts of pain.

A few weeks after that dinner at Sloan, I was introduced as the head of a sixty-person division of ESPN covering the NBA in digital and print media. Just as sitting was becoming an emotional trial, I would spend long hours in conference rooms, behind the steering wheel, crammed into airplane seats, and hunched over in front of a computer. My kids mocked me ruthlessly for it, but out of desperation, I became the guy stretching his hamstrings in the boarding area.

I might have felt bad for Phil Jackson at the 2009 Finals in Orlando; but by the 2014 Finals in Miami, I felt like Phil Jackson. As the Miami Heat battled the San Antonio Spurs, I battled a tricky hip and a trickier back. I slept poorly and stopped running. When I finally got an MRI in the middle of the Finals, I hated the feeling of slipping on surgical booties before climbing into the MRI tube; I just wasn't ready to be frail and medicalized. What I hated even more was the radiologist's grave tone as he called to tell me about a torn muscle in my pelvic floor, a torn labrum in my hip, a worrisome lower-back problem with a complex name, and (after I asked, joking, "Is there anything else?") pelvic bones that rubbed together in front, known as *osteitis pubis*. None was fatal to me as a person; each might be fatal to me as an athlete. I was recommended for back surgery, pelvic floor surgery, physical therapy, and whatever other medicalized hell might follow.

When you're hurt, a lot of the professional advice comes from the three horsemen of the free-movement apocalypse: radiology, surgery, and pharma. They circle injured athletes like vultures, eager to feast, it feels, on the carcass of your athletic life. They put you on a table or in a tube and tell you to lie still. Sometimes they cut you open and repair something, and then recommend milquetoast baby steps that fall well short of a full return to movement. The current regime might get you back to driving a car and sitting at a desk, but no one seems to care whether or not you get back to the bouncy movements that make life so fun. Had a flubbed half step in a half-marathon really ruined the second half of my life?

Knowing that research shows that back surgery comes with massive risks (and, probably, more surgical booties), I never considered it. But I could feel the vibrancy slip from my life. The Bulls might be able to get another Derrick Rose, but Derrick and I would keep our injured selves forever. I thought of all the friends and family mired in the aftermath of big injuries, and swore I would do whatever work was necessary to keep moving. But what *was* that work?

I got interested in the injury epidemic because I cared about the NBA and its players. As I tasted my own athletic fragility, it became clear how quietly and constantly athletic dreams shatter. The degree to which the world shrugs at that carnage feels dangerous. Was there a way to avoid the three horsemen? Could someone show me how to prevent injury in the first place?

After years of researching this question, I found myself in Santa Barbara, where I visited the Peak Performance Project and met Marcus

Elliott, MD. At P3 (as everyone calls it), Marcus and his team had developed a method that not only could have predicted Derrick Rose's injury, but might have prevented it from happening in the first place. You meet a lot of people in the world of sports who "have a method," but Marcus had something better: he had data.

Marcus tore his ACL at high school football practice in 1982. For months, Marcus mourned his college football dreams, but he emerged with a life plan that drives him still. To paraphrase the main character in Andy Weir's *The Martian*, Marcus resolved to science the shit out of sports injuries. He studied under globally known physiologists, surgeons, and exercise scientists on his way to a medical degree at Harvard. Marcus devised a program to prevent hamstring injuries among the New England Patriots; then the Seattle Mariners made him Major League Baseball's first director of sports science.

With each passing year, Marcus's mission clarified. "The big wins, the big impacts of medical history to this point, they're all related to prevention," says Marcus. "Like, *all* of them."

Take heart attacks. In the 1950s, cigarettes, red meat, mixed drinks, and TV-watching inertia were seen as part of a modern lifestyle. That heart attacks claimed lives by the millions was mostly seen as unrelated bad luck.

Then researchers examined the cardiac epidemic through the lens of the echocardiogram. This revealed the critical factor that had eluded doctors for millennia: arterial blood flow, which slows years before an acute crisis like cardiac arrest. Now doctors can see that slowdown early and prescribe diet, exercise, and medicine for their patients long before they are in severe danger. That gift to humanity is measured in millions of life years.

Sports injuries, Marcus says, remain in the "before" period of that story. Most of us move however we want—or barely at all—until something hurts. "If it gets bad enough, you'll go to an orthopedic surgeon," Marcus explained. "They have three options: inject, operate, or tell you to just stop doing activities you love." Phil Jackson didn't stop coaching because he lost his touch. Phil stopped coaching because his body couldn't take the rigors of travel, and no team would agree to his scheme to skip most road games in deference to his physical limitations. Our current setup is "a ridiculously bad model," Marcus says.

Instead, Marcus reveres bold athletic movement. His evidence-based conviction is that our bodies are designed better than any Tesla, iPhone, or fighter jet—and they're designed to *move*. His theory is that injuries are caused by physics, especially landing forces that, with a bit of biomechanical insight and training, can shift from destructive to productive.

In 2006 Marcus opened P3 and hired brilliant people. They sank force plates into the floor and hung 3D motion-capture cameras from the rafters. P3 began collecting "scalpel level" movement data from the explosive jumps, lateral slides, and landing impacts of Olympic volley-ball players, professional surfers, sprinters, decathletes, German soccer players, NASCAR pit crews, NFL players, a good percentage of Major League Baseball, and more than half the NBA.

In the beginning, Marcus gained clients because, as a Harvard medical grad, he appeared to be the most qualified trainer on earth. He also had freakish intuition—athletes were blown away by his ability to diagnose biomechanical issues with his naked eye. They came to him because they needed their bodies for their jobs. And agents, teams, and managers hired him to keep their players healthy to win games.

P3 still relies a little on Marcus's keen eye—he described my hip issues better than anyone after merely watching me walk around. But mostly, P3's program is built around data collection, machine learning, and rigorous biomechanical science. After thousands of motion-capture and force-plate scans, they may have the world's largest database of elite athletic movement, a forest of information bursting with seemingly magical insight.

As in cardiology, the breakthroughs in biomechanics come from seeing movement. Still images from the X-ray table or in the MRI tube miss so much! An MRI of Derrick Rose's knee the morning before he tore his ACL likely would have shown fantastic ligaments. But the MRI could not capture the forces that put his knee at risk with every landing.

Among medical professionals, ACL tears fit into the category of "catastrophic injuries" that threaten careers, require major surgery, and demand long recoveries. Sometimes we hurt ourselves while leaping, but P3 found that the "catastrophe" is very often a hard landing. Newtonian forces travel from the initial ground impact up into our bodies. If we don't channel those forces into strong soft tissue, they run amok.

In mechanical physics, the study of airborne objects is called *ballistics*, a word often associated with weaponry. But ballistics also play a critical role in biomechanics, specifically our understanding of the forces involved in leaving the ground and returning to earth. *Ballistic* is a term that comes up at P3 every day, if not every hour. Ballistic movements include running (but not walking), jumping, leaping . . . anything that gets you off the ground. The data that emerges from force plates are expressed in charts called *force-time curves*, where the giant spikes of dangerous physics almost all appear during landings. Big jumps, and especially landings, are to sports injuries as arterial blood flow is to heart attacks. That's where the action is. We can do for catastrophic sports injuries what medical science has already done for heart attacks: prevent them by the million.

A key P3 test has athletes step off an eighteen-inch box, land with each foot on its own force plate, and explode into the sky. Data from a single drop jump fill more spreadsheet columns than Excel permits. As the athletes land and walk away, the biomechanist can see the landing forces the athlete's body has endured. A particularly efficient mover might land with the force of twice her body weight—which is a lot for a body to manage. But it's nothing compared to P3's current record holder, who landed with the force of *nine times* his body weight. It doesn't take a Harvard medical degree to understand the destructive potential of catching a falling bison. Every volleyball or basketball game features countless hard landings.

A tempting response is avoidance: Lower your jumps to reduce forces. Cushion everything with squishier shoes. Or stop jumping entirely. This fear makes sense. Soldiers protect themselves against ballistic violence with Kevlar. But to P3, there's an essential difference between bullets, which we didn't evolve to absorb, and landing forces, which we did.

Our brilliant human physiology contains a robust system to man-

ZION WILLIAMSON

age ballistic forces: a triple-stack of strong joints—ankles, knees, and hips—each with range of motion, braking power, and complex supportive musculature. With symphonic timing, a well-trained set of legs can happily cope with almost any landing. In fact, for the best movers, the big forces of landing are *helpful*: their legs store the forces in their muscles like rubber bands, and use them to snap into the next move.

The problem isn't landing; it's landing *technique*. Even professional athletes have oddities of movement.

In Marcus's view, the best basketball player in the world right now should be Zion Williamson. When Zion jumps, he pushes on the ground about as hard as any human ever has. (In P3 data, Zion's the record holder.) The size and weight of an NFL linebacker, Zion's ballistics are truly extraordinary. "He doesn't have to tell you," Marcus says. "He shows you. It's better than language. It's more precise." Zion can *move*.

But Zion has barely played in the NBA. P3 has assessed Zion several times, but never trained him. Marcus has lost sleep over Zion. Bodies can take a lot, but land with just one bad angle and the leaning tower of your leg joints can easily Pisa. Newtons of force that wouldn't ruffle a quad or glute might mistakenly end up in a hamstring, a meniscus, or a small bone of the foot—all of which Zion has injured.

The biomechanists at P3 study which part of the foot touches down first, how the hips track, and whether or not the long bones of the leg rotate. They watch to see if the athlete's hips sway as they step off the box, the shin angle as ankles flex at the bottom of the jump, and the forces running through the left side versus the right. The data from that kind of assessment doesn't just pinpoint vulnerable joints, they also suggest the custom playbook to reduce any human body's particular injury risk.

Take ACL tears. To radiologists and surgeons, a torn ACL is a tragedy set on a stage just five millimeters wide. Researchers have published papers about injuries related to the size of the ligament, the quality of its tissue, and the width of the little canal where it lives deep in the knee. P3's view of ACL tears is literally bigger: while ACLs themselves inhabit the knee's interior, the movements that doom that tissue almost all take place in adjacent joints: the hip above, and the foot and ankle below.

Unlucky in the middle, knees catch hell. P3 finds you can go a long way toward preventing noncontact ACL tears by strengthening the muscles of the lower leg, landing with toes up, getting hips strong and mobile, and practicing bouncy plyometric exercises.

When there's trouble with arterial blood flow in the heart, the solution is not to stop using the heart; it's to get that blood moving more. The same is true of ballistic movements. With careful preparation, and in accordance with every body's own limitations, P3 encourages athletes to jump in almost every workout, starting with hops back and forth across a dowel, and building to the most aggressive jump training in sports, in order to hone the skill of good landing. The value of that work is confirmed by rigorous research outside P3, which shows, in one example, that interventions teaching young female athletes how to land reduce ACL injuries by an incredible 64 percent.

After deep learning about how any one athlete moves, P3's staff of biomechanists and trainers develop custom training programs, drawn from more than a thousand moves in different combinations. Some of P3's findings include an appreciation for crucial but underappreciated muscles like the soleus, gluteus medius, and psoas. There's also a lot of talk about new kinds of athleticism like braking, rotating, and, interestingly, moving with joy.

Marcus sees deep cultural value in the movement of Anthony Edwards—the modern Michael Jordan. Marcus winces when discussing boring elliptical machines and has a garage wall covered in the different longboards he uses to skate down mountain roads. Ballistic movements mean a lot to Marcus, and not only at work. He's forever seeking out cliffs to jump off, breaks to surf, and massive ramps to fly over on his snowboard. If his kids are bored, he challenges them to jump over obstacles like hurdles or rocks. If you're going mountain biking, he'll say to "get some air under the tires." Just as you're more likely to adhere to a diet if the food's delicious, you're more likely to keep moving if the movement's delightful.

The surprise of the P3 experience is that, more than teaching muscles, P3 trains brains. Language fluency can be drilled and improved. Movement works almost exactly the same parts of the mind. The linguistics system piggybacks on bigger and more robust networks that control human movement. Strength, range of motion, command, posture, spring, relaxation, and reaction times can all be improved just as verbs can be declined and vocabulary memorized. Fluency in movement might be the most useful fluency of all. "What do you call these kinds of moves?" a friend asked while attempting a P3 assignment, standing on one leg while rotating against the big force of a thick rubber band. "Brain exercise?"

P3 has an evidence-based case that the immobility and chronic pain that leech joy from so many athletic lives are neither mandatory nor inevitable.

I got a hint of this kind of thinking after my crisis in 2014. Once I got back home from the Finals, I began sweating through carefully managed and very difficult workouts led by kinesiologist Eileen Vazquez. I squatted with feet so carefully placed, deadlifted with my butt back and my shoulder blades pulled together, and heard intriguing phrases like "pull-ups are sneaky good for the pelvic floor." None of what we did was covered by insurance, or well understood outside those walls. But six weeks after we began, I had a holy moment as I rinsed a plate at the kitchen sink. I was pain-free.

I took it all as confirmation that the future of managing athletic bodies would be, should be, and must be different from the past. People with MRIs like mine commonly progress from surgery to immobility and disappointing decline—all the while elevating risks of mental health issues and opioid usage. It's hard to find experts who know much about getting you moving when you're in pain. My father was a brilliant doctor, but he admitted that neither his medical school education nor decades as a healer taught him much that mattered about the movements of soft tissues. When his back became painful, he began flying thousands of miles to see Eileen.

In the pandemic, I began no-showing for Eileen's Zoom workouts, put on a few pounds, and could almost physically feel my spine slipping into bad habits. As the pandemic ended, and I began flying back and forth to Santa Barbara to learn P3's secrets, the dark angel of lumbar crisis revisited, worse than ever. That made my life a living hell, but ultimately led to my throwing my body into a P3 assessment and pre-

scriptive workouts. I learned a ton about my own weaknesses, and boun-
tiful future weak points. I emerged with a personalized set of workouts
that have left me feeling another level of robust, and generally able to
do whatever I want—including lining up against Lance Armstrong in a
demanding Hyrox competition on a pier off Manhattan. I'm not sure I
have ever smiled more.

It's easy to see a revolution brewing. *Of course* big change is coming. The old system of shuffling around in pain is laughably archaic! We
can—and will—manage our bodies vastly better. The greatest thing we
can aspire to is the bounce and wiggle of real mobility.

More than anything, I've learned not to be afraid. Bodies are not
weak. They're brilliant, strong, and wired to move. Staying home is
killing us; Marcus wants us to get out and do the things we love. Real
delightful movement is healthy, necessary, and—with a little insight—
more realistic than we ever imagined.

# 1

......................

## MEDICINE

......................

*We were living in a time of almost*
*unbelievable ignorance about heart disease.*

—Paul Dudley White

DAVE CAME FROM the Navy SEALs, and rode a worn-out Pee Wee
Herman–type bicycle through terrible weather in the wrong gear.
Magna cum laude at Pitt, Albert had led a research project involving
bunny-rabbit bone grafts. Together with their Harvard Medical School
roommate, Marcus Elliott, they lived across the street from the Brookline Reservoir, where they swam, against the rules, even in the dead of
winter. Biking to campus on garage sale bikes became a ritual, a spiritual
practice, and an identity. In the worst winter storms, classmates cheered
as they pulled in, "then you'd rip off all your gear like Superman, and
underneath you're a doctor," says Albert.

One icy-cold morning, Marcus bellowed his "soo*wee*" war cry as he
ripped across the frozen surface of the Charles River. The conventional
wisdom is never to walk, let alone bike, on a frozen river, but Marcus
and Albert were not conventional.

"We took it on," explains Marcus. "I'm not claiming it was smart,
but that was a special thing."

Fast bikes crash hard. "Like a banana peel," remembers Marcus, of the river's sheet of ice, covered in a new blanket of snow. "The ground slipped out. I fell pretty hard."

In its tumble, Marcus's body wiped the snow clean off the river's surface. Face down, Marcus got his first good look at the surface. *Trouble*. He'd been expecting six inches of ice, hoping for ten. This looked like . . . two? "I could see leaves and things traveling in the river," Marcus says. "It put some fear into me."

The water ran close. And *fast*. Marcus remembers thinking that if he fell through, he'd have to grip the ice edge for dear life.

Crawl away? Run? Scream? Panic? Marcus . . . did physics. He remembers "some calculations." If he weighs 185 pounds and his bike is 30 and he crashes at 27 miles an hour, how much force would he generate? He considered the air temperature, water temperature, and thermal conduction. If the ice is two and a half inches thick, what force could it withstand?

Humans have been playing, exerting, and thrill-seeking essentially forever. Some of the oldest known human creations are cave paintings, and from Mongolia's Bayankhongor Province to France's Lascaux, they depict, among other things, sports.

And injury. Half a millennium before Christ, Hippocrates, the "Father of Medicine," said "health is at risk when exercise is at very high levels."

So, it might be a little surprising to learn that, while sports are ancient, sports medicine is, by comparison, a newborn. Taping ankles only became commonplace thanks to a man who died in 1986.

An extraordinary high school athlete, Augustus Thorndike made the papers in 1915 for dominating a high school track meet by winning the forty-yard dash, three-hundred-yard run, and twelve-pound shot put. Thorndike joined the Harvard football team but only played a short while before World War I drew him to the navy, after which he followed in his father's footsteps and became an orthopedic surgeon. Eventually, Thorndike was appointed chief of surgery for Harvard's health services, with residencies at seemingly every prominent institution in greater Boston.

In the early 1930s, a Harvard administrator asked Thorndike to look after the hockey team; other sports quickly followed. Thorndike served in two world wars and earned a Legion of Merit from the Pentagon— but it was his work on the football sidelines that made headlines.

Athletes today take for granted a level of care that Thorndike invented. He resolved that doctors, and not coaches, should determine who's fit to play. He arranged for an X-ray machine near the field. He banned any player who'd been knocked out cold three times. He told football players to grow their hair long for extra protection inside the helmet. Thorndike became, reportedly, the first doctor to design protective gear for football players, and wrote a pioneering book on athletic injuries, and another on taping and bandaging.

Perhaps Thorndike's biggest innovation, however, was to be present during the game. In 1940, *Time* magazine reported that when the Harvard football team took the field, a "heavyset, long-haired doctor stands on the sidelines, ready to rush on the field." Nearby, he positioned a dentist; a neurosurgeon; an ear, nose, and throat specialist; and an internal medicine doctor. When Thorndike ran out with his black doctor's bag, he could summon any of these specialists with hand signals. Eventually, he created a special class at Harvard Medical School for third-year surgical students, called Surgery 34. It typically comprised eight students who patrolled the sidelines of Harvard's other sports.

In one incident a football player named Charley Speyer took a shot to the head and began using nonsense hand signals. A Harvard teammate recognized the signals from years earlier, when they had been teammates at Choate boarding school. In other words, the collision had knocked his brain back a few years. After this, Thorndike recruited and empowered team captains to keep an eye on players, and to make it known if anyone seemed off.

Thorndike concocted a robust program to count and study injuries. *Time* describes one of his medical journal articles as "a meaty chronicle of sprains, strains, ruptures of the spinal column and spleen, fractures of the skull, collapses of the lungs" that "might persuade a casual reader that a sports doctor's life is a gory one."

In the waning years of his career, Thorndike shifted focus. The future would be less about running on the field with a black bag, and more about preventing injuries in the first place. In 1956, Thorndike published an article in the *Journal of the American Medical Association* that opened with the observation that nine players *died* during the 1955 football season, a problem he wanted to address with data collection. "A preventive program," he wrote, "must include a thorough preseason physical examination, with special emphasis on the history of prior injuries that the examinee might try to hide. There should then follow a period of physical conditioning under close medical observation, no matter what the sport or competition."

Paul Dudley White, MD, a Harvard contemporary of Thorndike's, was similarly upending cardiology. White's medical journey began in childhood, when his sister Dorothy lost her life to heart problems. After decades of mapping the underlying dynamics of heart attacks, White convinced the world that the best time to treat heart attacks was before they occurred.

Speaking at Harvard Medical School's graduation in 1956, White said cardiology had traditionally been completely irrational. Heart problems had been seen as acts of God, vengeance, or bad luck. In Roman times, Galen was the expert, but he worked largely with pig cadavers and made erroneous conclusions—for instance, that blood originated in the liver. Early practitioners puzzled for years over whether arteries connected to vessels. In 1707, the pope's physician, Joannis Maria Lancisi of Rome, felt compelled to write a whole book dispelling the myth that prominent Romans had been struck down by a spiteful deity.

Robust science began to arrive with that little beepy machine next to the hospital bed: the echocardiogram. Dutch doctor and physiologist Willem Einthoven won a Nobel Prize in 1924 for perfecting the machine. A decade prior, White had traveled by steamship to London to observe the emerging technology. Even though he served overseas in World War I, White conducted more than 20,000 echocardiograms between 1914 and 1931.

With this tool, cardiology could finally see that *movement was the thing*; the flow of blood (or not) through the dollhouse plumbing of the human heart determined whether the organ continued to pump. The machine had flicked the lights on in a dark room. White's thousand-page 1931 textbook *Heart Disease* quickly became a standard, and is still in use today. "Virtually singlehandedly," writes the *New York Times*, "Dr. White laid the foundation for today's American cardiology."

White made news by opening a bike path in Chicago and pedaling across the macadam of Martha's Vineyard to demonstrate the importance of exercise. "Death from a heart attack before the age of 80 is not God's will," White said, "it is man's will."

In August 1955, President Dwight D. Eisenhower flew to Colorado for a couple of months of work at the "White House West." One night he went to bed early, felt sick, and woke up with chest pains, which his doctor treated with painkillers. Later that day, an echocardiogram discovered a heart attack, and the Secret Service whisked the president to Fitzsimons Army Hospital.

Newspapers compared the ensuing uproar to Pearl Harbor. They summoned White, who arrived by military plane and examined the president in fine detail. With a flight waiting to take White back East, he lingered before the White House press corps. He had dedicated his life to spreading his message about heart health. Why waste the world's attention?

That day, White gave what, decades later, in the *New England Journal of Medicine*, Thomas H. Lee, MD, suggested was the most consequential press conference in history. "He spoke to the press, and by extension the public, as if they were medical students interested in understanding the heart and myocardial infarction. He explained how heart attacks were caused by the buildup of atherosclerosis in the coronary arteries and the formation of blood clots that blocked the vessels. He described how the damage was repaired the same way that cuts on the skin healed."

Family history, gender, age, and build probably all play a part, White said, "and there are a number of other factors that we are not sure about that are environmental, which include activity, occupation, stress and strain, diet, and customs, local customs, use of tobacco, alcohol, and so on . . . probably some of you in this room have coronary artery disease and don't know it." White became a media favorite. Press secretary Jim

Hagerty said that reporters told him it was the best press conference they'd ever attended.

White followed up with President Eisenhower every couple of weeks. Each visit produced front-page news.

"The editor of an antismoking newsletter wrote him that nothing published by any association had had impact comparable to what he had accomplished that day," writes Lee. "Heart attacks became less mysterious and less frightening to millions of Americans . . . White gave them the message that they could take steps to reduce their risk."

White's prescription for Eisenhower included riding his bicycle, which seemed outlandish at the time. In the years that followed, the medical profession made, as White put it, "adequate study" of the problem and determined the underlying factors of heart disease: high blood pressure, high low-density lipoprotein cholesterol, diabetes, smoking and secondhand smoke exposure, obesity, unhealthy diet, and physical inactivity. Treating heart attacks by preventing them became the standard of care, with doctors recommending improvements in diet and exercise, as well as blood tests for risk factors, anti-smoking efforts, and early diabetes intervention. At the time of Eisenhower's heart attack, Americans lived, on average, to sixty-eight. The death rate from heart disease was 588 per 100,000. Heart disease has fallen every decade since, to, the CDC says, 162 per 100,000 in 2019—a 72 percent decline. It's seen as one of the most successful medical interventions of all time, with the results measured in millions of life years. Americans now live, on average, a decade longer.

The names Thorndike and White were legendary at Harvard Medical School when Marcus and his roommates arrived. Their work had been transformative and brought glory to the Cambridge campus and Boston-area hospitals. But in Marcus's view, Harvard Medical School itself did not always reflect their maverick brilliance. He found the institution militaristic, hierarchical, and at times antithetical to much of what he wanted to do. "One week you're cutting open somebody's stomach," he says, "the next week you're in a psych ward, then after that you deliver a baby." All the while, Marcus dreamt of movement: mountaineering, surfing, or trail runs. If he had forty free minutes, he would bound up the hospital stairwell in scrubs. If his pager went off, he'd see patients while dripping sweat.

Harvard students get a big dose of medical history. But medical history, Marcus says, was mostly misadventure. "We just kind of stumble along in medicine," he says. "In Victorian England, it's well documented that the wealthy died at a much younger age than regular people, because the practitioners were applying too much medicine. Too much bleeding, too many leeches, too much arsenic." The practice of medicine has been around almost forever, but Marcus notes that only around World War II did the field show a measurable impact on health.

"I learned to be really grilled in the scientific method," says Marcus. "You have a hypothesis, you collect data around it, and you see which direction the data takes you." That's what White brought to cardiology, and what Thorndike thought was missing from sports. "The obvious thing that was missing was applying the scientific method to musculoskeletal health," Marcus says. "At that time, it was just not how it was done. And it's generally still not how it's done."

Just as White found clues in the movement of a heart years before a heart attack, Marcus wanted to examine the movement of ankles, knees, and hips in the years before an ACL tear, an Achilles rupture, or a blown-out back.

As he lay on the frozen Charles that morning, Marcus used physics to assess the impact forces. His calculations revealed what everyone else had known without calculations: it was dumb to ride a bike on that frozen river. Marcus and Albert padded gingerly to the river's edge, then rode the rest of the way to school on a bike path named after Paul Dudley White.

# 2

## FREE-RANGE

*Always be on the lookout for
the presence of wonder.*

—E. B. White

WHEN BERA LISLE WOODY died in 2003, her obituary said she
was born in Coal County, Oklahoma, in 1917, moved to Arkansas,
where her husband owned a general store, and worshipped at the
Church of Christ.

It did not mention that her son Gaius T. Woody did "peyote for five
days straight with this old Mexican guy down in the desert," according
to her grandson Marcus. "Totally changed his life course."

"Next thing you know," Marcus says, "he's living in Big Sur, out of
a yurt, and carving pieces of wood that he found. He became a beat-
nik, hanging out in San Francisco." Gaius married a woman named
Rochelle; their three children grew up as free as wild ponies.

Many kids in the 1970s roamed around Bolinas, a hidden surfing
town an hour north of San Francisco not many years or miles removed
from the acid tests of Haight-Ashbury. But Leah, Marcus, and Jenna
had it better. With their house rented to tourists for extra cash, the
family spent entire summers under the scrub trees on the hard sand

on the edge of town, living in a Steinbeck novel.
They cooked over wood flames, fell asleep in cots,
breathed noisy sea air, and woke to sunbeams.

Sometimes Gaius would put Marcus on
his shoulders and amble out onto the reef with
cheap fishing rods. They knew how to keep their
shadows from frightening the rockfish and when
the halibut would bite. "We'd outfish all these rich
guys with their killer equipment," remembers Marcus. He liked returning with dinner. He loved out-
thinking the rich guys.

Then Gaius met Russel V. A. Lee. It turned out
that, in addition to being a forward-thinking
doctor, a founder of the Stanford Medical
School, and a big-time real estate investor, Lee was an art collector. He
offered the Elliotts rent-free living on property he owned in the Dry Creek
Valley, which cuts through Sonoma County's stunning Mayacamas Mountains with features like Frog Woman Rock and Panther Ridge. Instead of
money, Lee wanted a right of first refusal to purchase Gaius's work.

But Gaius didn't take the deal. It didn't sit right to live in someone's
house without paying. So, he bartered: the Elliotts would pay fifty dollars a month for the six-hundred-acre property in Preston.

Now the Elliott-Woody children (Marcus would later drop the
"Woody") roamed over six hundred acres of black walnut, chokecherry,
laurel, and dogwood. The land had a house, a spring, a giant fig tree, a
lake, and a barn for sculpture. Marcus taught himself to throw walnuts
so they'd hit any branch you pointed at, how to ride a contentious pig,
how to solo camp on a rocky ridge a day's walk away, and how to launch
his body from the branches of the fig tree into the cold water below. On
hot days, Marcus would gallop on a horse with his husky, Casey, trailing
them across the crunchy dry grass, to the lake a mile away. They barely
slowed at the lake's edge.

Marcus built things with Gaius. "I had my dad's hammer. It's
super-heavy and I'm trying not to smash my fingers. I'm scared." As
they built a treehouse, Marcus was reluctant. "Somehow it's stuck

with me so clearly, but slowly he said, 'Marco, Marco, don't picture hitting your hands.'"

Anticipation can stifle free movement. "See the nail?" Gaius said. "Don't think about it. Just zen it." Marcus still has that hammer, and that idea. *Zen* might be tough to define as a verb, but it was lasting advice. Getting deep into your thoughts doesn't help you hit a nail, backflip off a bridge over the Russian River, or hammer dunk in an NBA game.

At ten, a friend's mom signed Marcus up for Little League baseball. Marcus had somehow lived an American decade without seeing a single baseball game. When they asked what position he played, he said "pitcher" because it was the only one he could name.

Marcus liked the feel of the ball; his teammates loved that he could throw it hard, even if it took a few games to dial in the rules. Soon he starred on an undefeated team, and earned a call up to the older league. Sports became a huge part of his life: baseball, football, or just about any other outdoor thing you wanted him to do. Drives to practice and uniforms and cleats cut into his time with the horse and the dog and the pig.

Marcus's parents encouraged him. "I started to study econ to be an investment banker. I liked the idea of wearing a three-piece suit. My dad's like, 'Oh, that's great, you'd be a great investment banker.'" Marcus found chemistry incredibly difficult, but had some idea about becoming a chemical engineer. "I sucked at it. And it's like, 'You'd be a great engineer, Marco.'"

Or a football player. Marcus spent the afternoon of his seventeenth birthday at Cloverdale High football practice. "I'd just done this drill with this kid, it was, like, a one-on-one tackling drill," he remembers. "I juked him really hard. He was a big kid. He was bigger than me." By then Marcus was about six foot two and 185 pounds.

They did the same thing again, and Marcus had an even better move. Marcus played free safety and wide receiver and had college hopes. "I made, like, a shimmy-shimmy and went straight through him. Just like straight through him." He demonstrates the movement. Arms out, palms up, shoulders and hips working.

The big kid crumpled toward the ground, but caught Marcus's leg on the way. First, his falling shoulder "pushed in" Marcus's right leg. The force hyperextended Marcus's knee, which pinned him in place.

"I was just standing there with the ball," Marcus says. Even then, Marcus was a nerd about how the body works. There are four major knee ligaments, thought Marcus, as his teammate's falling body began to rotate through his joint. Marcus prefers not to remember this moment, but adds, "I remember exactly how it sounded."

Around four o'clock on that beautiful end-of-summer afternoon, seventeen years of free movement came to an end.

The physical break caused a social one. "Everyone was showing up in my hospital room. I had flowers all over. It was full!" Marcus remembers. Within a week, his coach had found another player to throw the ball to—one of Marcus's best friends. Before long, Marcus's girlfriend was dating that guy.

"Then I was kind of forgotten," Marcus says. "Like the world went on spinning, and I wasn't spinning with it anymore. I wasn't talking to my classmates. I missed four months of school. I had a full-length cast on for four and a half months, up to my thigh." He didn't go to the lake, play with the pig, nor explore the six hundred acres. "I was depressed. I didn't know what my life was gonna be. That's the truth. Like, it felt like, in a seventeen-year-old way, that my life is over," Marcus says. "I had something that was so precious to me taken away. And . . . and it hurt so much for so long. There's got to be a better way."

Marcus read the medical journal *Medicine & Science in Sports & Exercise*. And he started thinking. If there was a technique to shimmy past a tackler, to catch a fish, or to throw a pitch, there must be a technique to prevent this hell.

"When I came out of it, I was looking at this same whiteboard that was empty, and saying, 'Wow.' Sort of, also: endless potential. 'I can go *anywhere*.'"

He had a project. He would study the human body and learn how to prevent injuries. Marcus says, "It's very cliché to say, you know, 'I'm better for it,' but I really am. I worked so much harder. I was so much clearer about where I was going."

His parents told him he'd be great.

# 3

....................

# WICKED

....................

*In the most devilishly wicked learning
environments, experience will
reinforce the exact wrong lessons.*

—David Epstein

BY THE TIME Marcus got his head together, he had missed all the college application deadlines. His mom found a summer program at Berkeley that served as a stepping stone to the University of California, Santa Barbara, just a few hours' drive from home and, thankfully, surrounded by a swath of preserved wildlands roughly the size of Massachusetts.

Marcus's parents handed him what they had saved for college. The three hundred dollars fit easily into Marcus's jeans pocket.

When he got to Santa Barbara, Marcus applied for a dozen jobs, was offered five, and took three. "I worked at a liquor store," Marcus says. "I worked at Straw Hat Pizza, and the New York Bagel Factory, making bagels at five in the morning." After a feeling-out period, he quit the pizza gig and was left with days bookended by $3.35 an hour making bagels and $3.65 selling liquor.

Time and money were tight. Marcus was on three sports teams. He attended classes. And he needed to find a way to solve the complicated scientific mystery of sports injuries.

"There's not, like, an easy route to make that your life's work," Marcus says. When he left home in the late summer of 1984, there was no field of study dedicated to injury prevention. Marcus endeavored to catch a fish that had never been caught, and he didn't know where to cast his line.

About that time, a researcher at the University of Chicago named Robin Hogarth was examining how people learn. As popularized by David Epstein in *Range*, Hogarth suggested that games with clear rules and hard edges, like chess and darts, make "kind" learning environments—the dynamics of the thing become clearer as you play. Other problems, though, are "wicked" and packed with misdirection. Some problems, Hogarth says, are as complex as "Martian tennis"— people are playing all around you, but the rules are indecipherable and change all the time. Hogarth and his coauthors opened a 2015 article with an anecdote from a New York City doctor a century prior, who had a knack for finding typhoid cases: "His clinical technique included palpating patients' tongues before making his pessimistic forecasts. Unfortunately, he was invariably correct." We later learn that the doctor had typhoid, and was giving it to patients by touching their tongues.

Sports injuries have long been treated like simple problems: something would break, you'd find it on a scan, and then address it. But Marcus wanted to do something vastly more complex: to see the injury *before* it happened. There wasn't even a clear field of academic study that might capture it. How might hidden trouble signal itself? In blood values? Diet? Sleep patterns? Neuroscience? Physical therapy? Athletic training? Pick wrong, and Marcus might spend a decade down the rabbit hole of one field, only to learn that he needed to look elsewhere.

Where on campus would Marcus find a mentor? The answer turned out to be the men's room. Before he had taken his first undergraduate class, Marcus stepped up to a UC Santa Barbara urinal. As he recalls, a particularly short man began peeing next to him. "I swear I'm not in the habit of talking to strangers at urinals," says Marcus. But there was something about that guy; Marcus struck up a conversation. "And he asked me what I was interested in. And I told him. He goes, 'Well, that's what I do.'"

That's how Marcus met legendary sports physiologist Steven Horvath. Marcus loved listening to Horvath, who often gave Marcus rides in his convertible Cadillac, smoking cigars and dropping pearls of knowledge about everything from interracial relationships (Horvath was in one) to surviving extreme cold (Horvath was a global expert). There was something about Horvath's approach that fires Marcus up to this day. "Like, he was just curious as fuck," says Marcus, "and that almost gave me a green light to be as curious as I already was."

Horvath began his career exploring limits at the controversial Harvard Fatigue Laboratory, where physiologists, biochemists, psychologists, biologists, physicians, sociologists, and anthropologists subjected research subjects to treadmills, a "climatic room," an altitude chamber, and a room for animal experiments.

In World War II, the military commissioned the Fatigue Lab to look into questions like how soldiers acclimate to hot climates. That's when Horvath, a Harvard Medical School student and research associate at the time, made his mark. Hot weather had been demonstrated to inspire changes that could feel taxing, like higher blood volumes, increased sweat, and decreased urination. But the team's big finding was that the brain learned and adjusted. While getting acclimated to heat, the body's ability to regulate salt can get out of whack—increased sweating, in that phase, can deplete the mineral. But also, commonly, the body would become stingy, so that sweat would contain far less salt than normal, which can cause its own complications. Horvath's team found that the fix was to trust the body's wildly adaptable nervous system to work it out over time. "A man's ability to do prolonged physical work in the heat," they concluded, "improves markedly."

The result fit a theme. The Harvard Fatigue Lab found that fatigue seemed to have a little to do with the body and a lot to do with the brain. A well-trained athlete is sophisticated in achieving what they called a "biochemical steady state." There are limits to our performance, but they are tough to find in our physical construction.

After the war, the lab closed. Horvath ran a research institute at Iowa State before moving west in 1961 to found the Institute of Environmental Stress at UC Santa Barbara, where they did things like ask sixty-two school children to walk around a track in Nevada for an hour in 108-degree heat. By the time Marcus met him, Horvath had become a leading researcher into the effects of airborne pollution. Horvath found that pollutants from cars and power plants—especially ozone—presented grave health concerns. In a 1976 article, Horvath found that "ozone had a marked effect on the lung, causing a reduction in vital capacity." UCSB students had worked out on treadmills while exposed to varying doses of ozone, at different temperatures. Many complained of dry throat, headache, fatigue, or cough—and often struggled to take a full breath. During ozone exposure, Horvath found, lung vital capacity declined an average of 350 milliliters, which effectively gave healthy young athletes the lungs of someone two decades older.

The summer that Marcus arrived in Santa Barbara, the Olympics were just down the coast in Los Angeles. Horvath wrote that athletes would be performing under stress from heat and "the presence of an extremely potent photochemical smog that may have as adverse an effect on certain performances as did high altitude during the Mexico City Olympiad."

Horvath's lessons felt like a secret history of sports performance—the athletes may be top notch, but what about the air? If Marcus had more time, he'd explore it all. Marcus majored in biochemistry and shoehorned as many independent research projects into his schedule as possible. "Just big white-paper studies," he says, "where I'd go in and just study the hell out of some subject I was interested in. I'd read every academic journal I could find: VO2 max, lactate threshold, heart rate."

It's very rare for anyone to be on three college sports teams, but Marcus competed in track, cross-country, and cycling. If you had to pick a social stereotype for him, "jock" would have been it—unless you hung out with him on the weekends. Marcus remembers being so steeped in research that "I'd cite these studies from Weinberger in 1973." Marcus was a hard-liner in both camps—the nerdiest jock, the jockiest nerd.

Those two things, to Marcus, had come to feel surprisingly similar in his post–ACL tear life. At football practice, fifty people had motivated him by cajoling and hooting and cracking helmets. Crowds filled the stands on game day. Running, biking, and swimming were entirely different. "You're always training and racing in the shadows," Marcus says. "Nobody gives a fuck, nobody cheers, nobody cares!" The mental experience of cross-country training was built on self-talk, on curiosity, on ongoing problem solving—much like academic research. "That's an amazing driver, an engine design for life," Marcus says. "Studying in college is just like preparing for a triathlon. Nobody cared. You have to make a lot of choices to work when other people make other choices. This is the period when I was like, 'These are the real sports. Football, baseball . . . that stuff is soft.'"

And then, one morning at the bagel shop, he added another layer to his identity. As he dug into the cream cheese, a customer asked if he had ever considered modeling. "'I think you'd be great,'" Marcus remembers her saying, "and she gave me her card." Before long he was at an agency downtown, where they said promising things. *They would make him a book!*

But first, they'd need thousands of dollars for modeling classes. "I'm like, 'What the fuck?' This is the opposite of what I'm trying to do here," Marcus says. "On my way out, this guy was walking in, and he said, 'You'd be perfect.'" The guy wanted to know what Marcus would charge for a couple days' modeling work. Marcus made up a number and spent the following weekend making sixty dollars an hour. Soon, he stopped selling bagels. The money bought him precious time to learn about bodies.

Marcus applied his scientific knowledge as much as possible, including to how he breathed. "I just saw my body as a machine to process oxygen," Marcus says, and he learned the freakish power of a good exhale. A shallow breath brings air mostly into the esophagus, where there is no lung tissue to process oxygen. The air that matters gets into the lungs. So, Marcus began wringing air from his lungs with vicious exhales. To this day, you might hear loud bear chuffs from Marcus running uphill. Once you get the spent air out, Marcus points out, "you barely have to work to inhale. The lungs take their own shape again naturally."

As he got more into triathlons, Marcus began early-morning workouts with a swim coach and assigned himself miles of open-water solo swims. When his right shoulder got sore, his coach pointed out that his right hand crossed the midline of his body when he reached forward. When he put his hand where the coach wanted it, a centimeter to the right of the midline, it felt like it was three feet out to his side. But when he swam that way, his pain melted away.

This was interesting. Marcus didn't totally understand what was happening, but it felt like evidence that injuries signal themselves before they arrive. His shoulder had been vulnerable. Then he tweaked the movement, and the problem evaporated. It was small confirmation of Marcus's big idea, that there were ways to make athletic bodies more robust.

Marcus had been reading the work of Tim Noakes. As a rower at the University of Cape Town in the 1970s, Noakes was tricked. The coach told them to do six five-hundred-meter repeats. Then, surprise! The sneaky coach assigned another five hundred, and another, and another, and another, for a total of ten.

Studying to become a medical doctor, Noakes knew the research of the day declared that athletes stopped competing because muscles were at full work capacity. He felt that way himself. The sixth repeat had been all out.

And yet, Noakes wrote in his memoir *Challenging Beliefs*, his team's boat moved faster still in the seventh, eighth, ninth, and tenth efforts. No one could explain that, so Noakes thought he'd give it a try. "When I started my career, my lab had no money," Noakes later told *Outside* magazine. "All we could buy was a rectal thermometer and a weighing machine, a scale. All we did was go around measuring people's temperature and weight after races.

"In fact, we borrowed the rectal thermometer."

Noakes quickly began upending assumptions. A prominent colleague theorized that distance runners couldn't have heart disease. Noakes analyzed the corpses of long-distance runners and found they absolutely *could*. Just as the sport drink industry began to push hydration, Noakes showed that humans had invented a new way to die: overhydration, which diluted salt in the blood.

And in 1996, Noakes published his most important work, on what he called the "central governor." Noakes found that "fatigue is an emotion," a refinement and advancement of ideas explored decades earlier at the Harvard Fatigue Laboratory.

Noakes argued, essentially, that in his sixth repeat that day in rowing practice, he had felt fatigue long before his body was depleted of oxygen, glycogen, or any other vital resource—because instead of finding a hard physiological limit, he'd felt a subconscious signal, from his brain, to slow down. Noakes posited that fatigue was the product of the brain, which acted like a "central governor," to keep athletes from dying from overexertion or overheating.

It's an empowering idea. In *The Lore of Running*, Noakes writes about how "fatigue drives us back into ourselves, into those secluded parts of our souls that we discover only under times of such duress and from which we emerge with a clearer perspective of the people we truly are."

One of Marcus's first triathlons was in Santa Barbara, for which he wore a mere Speedo as he ripped along on his bicycle at forty-five miles an hour. He remembers being "maybe second or third overall" when a tire lost its grip, sending him sledding, as he recalls, "so far" along the asphalt on his backside.

Competitors shot past. Marcus stood and knew three things: 1) it would help nothing to look back there; 2) the pain was rooted in his emotions; 3) the research shows that human bodies are alarmingly resilient. Marcus pedaled to the changeover as fast as he could, then popped on his shoes for a ten-kilometer run, winding along waterfront palms.

The blood pooled in his sneaker. Step, squish, step. He overtook one runner, then the next. Marcus won back positioning; they won a close-up of his backside. "Other competitors," Marcus remembers, "were like, 'Stop!' They were horrified. They wanted to help me."

A full-length triathlon, at the speed of a podium finisher, tests a body's ability to process oxygen, in through the lungs and out to the muscles. The vehicle of that transfer is blood, which is why losing a lot taxes an athlete. (One method of "blood doping" in the Tour de France is to withdraw a pint of blood months before the race, which reportedly makes training arduous. Once their bodies replace the missing blood, riders return to feeling normal. Then they reinfuse the extra pint during the race, feel tireless, and, in cycling's doping era, sometimes win the Tour de France.)

Marcus didn't stop for the blood loss, though. He would run through negotiable emotional hurdles to see if he could find a hard physical one. Marcus feels he had a motivational advantage from his ACL injury: to him, dropping out also hurt. "It was really all set up," Marcus says, "after having this injury and the depression, knowing what a lost opportunity feels like."

At the finish line, Marcus didn't win, but he did feel like a butterfly shedding his cocoon. What a revelation: he could keep going even after powerful feelings told him to stop. So much of his training had been about times and weights and heart rates—but a tango with the impossible packed a far bigger punch. "I just felt," Marcus says, "so heroic."

Researchers following in Noakes's footsteps began exploring the ways the brain fools us. In his 2018 book *Endure*, Alex Hutchinson delves into what are now known as *belief effects*. One example: People made fun of Michael Jordan for his superstition of wearing his University of North Carolina shorts under his NBA uniform. But at the same time, research showed that saying, "Here is your ball. So far, it has turned out to be a lucky ball," got people to putt better than saying, "This is the ball that everyone has used so far."

Another study showed that if you race on a treadmill against a virtual reality representation of yourself, you'll keep up with that character—even if the researchers cheat and program it to go faster than you ever have. (If it goes *a lot* faster, though, you'll shatter.) But the implication is that it matters, a lot, how fast you *believe* you can go.

"For athletes, the simplest way of acquiring justified true belief about your capabilities is to test them: whatever you've done before, you can do again, plus a little more," writes Hutchinson of Noakes's work. Noakes also suggested that high-intensity training, in particular, could rejigger the central governor—sprint training, for instance, affected the central governor more than long runs.

Marcus found race-day strategy here, too. Noakes wrote, "I suspect that dominant runners intimidate the central governors of their competitors, forcing the intimidated governors to restrict the performances of their owners." Blow past people, and let their own central governors click on and slow them down.

Asked to identify the toughest workouts of his life, Marcus answers "the Death Rides." A pioneering coach of elite US cycling, Eddie Bory-sewicz, ran a team based in Escondido, California, that would change names several times before eventually becoming the notorious US Postal Service team. Professionals tuning up for the Tour de France, hard-core triathletes, college racers, and other assorted maniacs rolled north out of San Diego every Wednesday morning.

Marcus spent Tuesdays at his girlfriend's in Encinitas, and waited on the wide shoulder, early on Wednesdays, for the group to appear. "For the first twenty miles or so, you'd roll along at twenty-three miles an hour," says Marcus, who remembers that pace as manageable. "Then we'd get to Camp Pendleton and turn uphill, and it was just punishing each other." Sometimes Marcus let the leaders go. Sometimes he hung with the lead pack to the top. "If you were one of the ones dishing out the punishment . . ." Marcus peaks his eyebrows, cocks his head a bit to one side, and exhales with cheeks puffed out. *Big deal.*

In some ways, Marcus fit perfectly on a Death Ride, rolling with his outdoor tribe, chuffing uphill. But also, he was younger than almost everyone, and more academic—his head bursting with Horvath and Noakes. Marcus slid into a kind of trainer side hustle, advising the Subaru-Montgomery track cycling team and, for a time, legendary tri-athlete Mark Allen.

Harvard Medical School accepted Marcus, even though—it's Marcus's shameful secret—to this day he remains a few credits shy of his UCSB biochemistry degree. He deferred his medical school admission to model, lifeguard, live in a VW bus, and travel the West, competing in triathlons. He was sponsored by the likes of Hobie sunglasses against competitors like Lance Armstrong.

Near the end of Harvard Medical School, Marcus arranged what the school calls a clerkship, essentially an internship, with Noakes in Cape Town. There, Marcus found himself sitting outside, often with a stunning view, talking about bodies with brilliant people like physiologist John Hawley and Louise Burke—later the head of nutrition at the Australian Institute of Sport.

A theme emerged: one-size-fits-all remedies are almost always wrong. Coaches tended to put one workout on the wall; there was a lot of talk about what athletes should eat or what shoes they should wear. And yet, we all know, from standard medical care with its blood tests and X-rays, that we're each deficient and strong in different ways that should inform the remedy.

In South Africa, Marcus made a research project of race-day nutrition. One decade, athletes were told carbs were all that mattered, then came the protein years, followed by a decade of obsessing over designer fats. Each theory suggested there'd be a simple, bright-line answer for everyone.

During this time, fat was seen as the worst possible nutrient. Marcus ran a test on peanuts, with mixed results, but he noticed that some of the test subjects looked and reported feeling peppier with a bit of fat on board. It fit many signals suggesting that athletes would generally do better ignoring extreme nutrition advice, eating, Marcus says, "like your grandparents ate," with plenty of vegetables, some protein and fat, and not a lot of processed food. Hard-and-fast rules beyond that fared poorly. "We can't get the *macro*nutrients straight!" Marcus says. "That's not the complex stuff. That's like the ABCs of nutrition." To this day, Marcus lets hunger be his guide, which, he says, makes him one of the only athletic people in his Santa Barbara social circle who still puts cow's milk in tea and doesn't shrink from a wedge of brie. "That feels right to me," he says, adding, "I *wish* science had all the answers."

The 1996 World Athletics Cross Country Championships had brought many of the world's best runners to South Africa, eleven of whom visited Noakes's lab. Kenyan and Ethiopian runners ended up winning eighteen of the twenty-four medals at the championship, including every gold. Over two intensive days, they were assessed in every imaginable manner, from the size of their thighs to the contents of their blood. The simple finding was that the Black East African runners had no obvious physiological advantages (not VO2 max, not muscle fiber type composition, not muscle strength) over the other runners.

Nevertheless, they demonstrated "superior fatigue resistance" in the form of lower blood lactate concentration after heavy doses of running.

Some runners roll along flat and low, Marcus says, "like a tractor," while others bob up and down. To Marcus's eyes, the best runners from East Africa were "so much more bouncy than the white runners." He noticed how their feet struck the ground, and how the force of landing passed quickly up to the big muscles of the hips. The research was about fatigue resistance, but it gave Marcus ideas about springiness and injury prevention. "Everything in my brain," Marcus says, "from my experience in sport, told me that was safer."

In these years, Marcus's side hustle as a trainer evolved. He ran speed and agility clinics alongside Canadian national team coach Brent McFarlane and advised Olympic runners, including world-record sprinter Asafa Powell. A lot of his work was in preventing injuries, particularly in hamstrings. Before he left for South Africa, Marcus had gotten a phone call about that work from orthopedic surgeon Bertram Zarins.

"Zarins was an important figure," says Marcus, "because he was one of the top three sports docs in the country. By name, by recognition, by publication. Lead doctor for the Olympics. Team doctor for all the pro sports teams in Boston." They arranged a meeting that, on the wings of enthusiastic conversation, stretched to three hours.

Marcus remembers Zarins telling him about NFL players coming to camp with known back injuries, and having the strength coach welcome them with maximum-weight squat testing. Marcus remembers Zarins saying, "We had two guys blow up in testing." A lot of people assume everything that happens in elite sports is elite. But Marcus says Zarins "knows how dumb it is, because he lives there."

They resolved that, when Marcus returned from South Africa, he would join Zarins at the Patriots for a special project. After dozens of injuries, especially to star receiver Terry Glenn, the team's billionaire chairman and CEO, Robert Kraft, demanded that the team get better at hamstrings.

Running across the back of the hip and knee, the three main muscles of the hamstring are twitchy, generally brilliant, and particularly tested as a running foot hits the ground. Just as the hamstrings straighten the hip pushing the leg out behind, they also work to bend the knee, and then they catch forces coming up from landing. This raises questions about the tensile strength of muscles. "They tend to get weaker as they get stretched more, but there's an area where they get weaker faster," explains Marcus, who notes that a typical hamstring injury occurs in full sprint. "You're in a super-stretched position, which makes the muscle vulnerable, and you're not very strong, and you have all this weight coming down and wanting to flex your hip and maybe to extend your knee, and both of those things stretch your hamstring."

Marcus was years into hamstring research. Clearly, strength mattered. A buzzy 0.6 ratio had caught on in the industry. Hamstring injuries, the thinking went, could be reduced by making sure that the hamstring was at least 60 percent (or 0.6) as strong as the quadriceps. Sit on your butt, see how much weight you can push by straightening your legs against resistance, and that's your quad measure. Then flip to your tummy, and curl your heels toward your butt, and that's the hamstring measure. Tidy and clean, measurable in any school gym or physical therapist's office, the 0.6 ratio had been around since the time of Eisenhower's heart attack. The ratio came from studies that explored the differences between athletes who had injured their hamstrings with those who had not; the healthy people had hamstrings that seemed stronger compared to their quadriceps.

Marcus says it might be "better to infer that hamstrings that are injured remain chronically weak." Under scrutiny, the 0.6 ratio fell apart. A 2002 survey of all available hamstring research concluded that "the cause of hamstring injuries is still unclear." Longitudinal studies found little or no correlation between the 0.6 ratio and injury risk. The whole concept had issues: mostly, the hamstring seldom operates alone—it's in the middle of the "posterior chain" of muscles, running up and down the body. Assessing it in isolation missed a lot of nuance. On top of that, they measured its strength while shortening, bending the leg at the knee. But hamstring injuries tend to happen with the leg moving the opposite way, extending out in front.

Yet the 0.6 ratio lingers in gyms and physical therapists' offices nonetheless, arguably because it papers over a very confusing reality.

"We're searching for answers," says Marcus, "which is the human condition. So, they found some ratio they liked."

Marcus pored over the data. "Every hamstring injury that happened in the NFL, I had sliced and diced," he says. Marcus wanted to know if hamstring injuries happened at certain times of year, at certain positions, or to athletes with certain biomechanical signatures or medical histories.

Marcus found evidence the answer to every one of those questions was "yes." NFL hamstring injuries usually happened in preseason, at positions like receiver and cornerback that sprint in the open field. And four kinds of players were especially vulnerable: those who had weak hamstrings, who had prior hamstring issues, whose sprinting form degraded as they ran farther, and especially those who sprinted with a posterior tilt to their hips.

Picture the pelvis as a bowl of cereal. Ideally, it's level, spilling no milk. Marcus saw that players who had the bowl tipped backward had a higher likelihood of stressed hamstrings. "When they sprint," Marcus says, "they tend to overstride with this posterior pelvic tilt." Effectively, their hips aimed their legs forward, so the feet flew out in front. "And a lot of them," Marcus continues, "when they contact the ground, on initial contact, they flex their hips a little bit more. So, they don't immediately go into pulling. They actually sink into that a little more. Mechanically, that should put a significant stretch on the hamstring." Marcus is convinced this posterior tilt is a big factor. Most hamstring injuries occur in fantastically powerful athletes; when more human

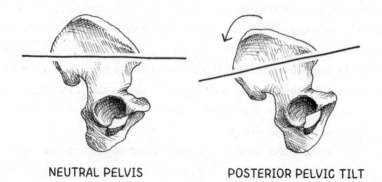

NEUTRAL PELVIS          POSTERIOR PELVIC TILT

people injure hamstrings, Marcus finds the reason is almost always poor pelvic control.

"The thing about hamstrings," Marcus says, "is that you don't want to have a second one. The second one is so much worse." It's not in the literature, but what Marcus observed was that a first hamstring injury did two things: it suggested biomechanical issues, and the injury itself left a legacy. Often, there would be a knot of scar tissue, and muscular adhesion, where stressed collagen fibers stick to adjacent tissue, all of which makes a muscle less stretchy. On top of that, it's common that the nervous system would get antsy. "When you have some protection going on," Marcus says, "you don't let it relax completely, it doesn't stretch completely, you lose elasticity."

Zarins helped Marcus get funding through the NFL Medical Committee. They gathered a decade of data from across the NFL and found 1,716 strains. "More than half (51.3%) of hamstring strains occurred during the 7-week preseason," Zarins and Marcus would later write, with colleagues. "The most commonly injured positions were the defensive secondary, accounting for 23.1% of the injuries; the wide receivers, accounting for 20.8%; and special teams, constituting 13.0% of the injuries in the study."

Marcus had to adjust to a culture that wasn't as curious as Harvard Medical School or Cape Town. Marcus remembers Zarins blanching when he pitched that someone should study the long-term effects of anabolic steroid use—which was evident, a health concern, and not well researched. "I didn't understand the environment," says Marcus. "I wouldn't do that now."

The Patriots' strength and conditioning coach, Johnny Parker, Marcus says, "didn't like me. Honestly, it was my first experience with that, of feeling that I was clearly bringing something right, contributing, helping the team, and it didn't matter." Parker had been an NFL strength coach for decades; Marcus was a twenty-seven-year-old with opinions. "I kind of get it now," Marcus says.

But people across the Patriots organization jumped on the hamstring project. In a burst of activity before the end of the season, they measured every player's hamstring strength, charted their positions, and did video analysis of sprinting form. Every player earned a hamstring risk score, which was higher if you had weak or asymmetrical hamstrings, poor sprint mechanics, poor sprint endurance, or played at a high-risk position like defensive back.

Then Marcus spent much of the offseason visiting the high-risk players to deliver personalized programs designed to change their own unique patterns of movement to reduce the risk of injury. Marcus was bringing a teaspoon of White's preventive cardiology to the Patriots' hamstrings.

"We go from a mean of twenty-three hamstrings—22.9 or something—over the previous five years," says Marcus, "to three." The implications were massive. In 2019, the Associated Press estimated that NFL teams spend more than half a billion dollars a year on injured players, most of whom are dealing with lower extremity injuries—to say nothing of the multifaceted benefits to players and teams of health. Marcus felt they'd cracked the code, and was quoted in *Sports Illustrated* as saying, "If teams implement reasonably smart hamstring injury prevention programs, these injuries can largely be prevented."

Marcus was fired up to publish the results, but Zarins delivered crushing news. The Patriots, feeling they had a competitive advantage, wouldn't allow it. Everything would stay secret. "My first thought," says Marcus, "was 'Fuck that.'"

Marcus did find a partner in preventing injuries: Patriots trainer Ron O'Neil, whom Marcus calls "one of the best athletic trainers I've ever worked with." But the more cautious approach to injuries that appealed to O'Neil sometimes meant holding players out of games, which was at odds with new head coach Bill Belichick. "His model," Marcus says of the coach, "was 'get him out there.'" A February 2002 Super Bowl victory gave Belichick enough organizational leverage to make changes. Weeks later, the team announced that, after eighteen years, O'Neil would leave the Patriots to run a new sports therapy and rehabilitation center.

Marcus began to wonder if the NFL was the right place for him. The league had brilliant people like Zarins, incredible athletes, and, at least when Robert Kraft insisted upon it, a willingness to innovate.

They had demonstrated, with the hamstring project, that some bodies are more robust than others, and that there's useful work to be done to injury-proof elite athletes. One of the most tenacious enemies Marcus encountered on his quest to prevent injuries was a lack of curiosity, and the shrug as someone says, "Injuries happen."

Pro football also had, in the spirit of *Moneyball*, old-timers who just weren't all that curious. Marcus remembers the moment he lost faith in the NFL. He was in the stands with the scouts at the predraft combine in Indianapolis. Marcus watched the drills with a head full of science, obsessing over the implications of pelvic tilts, bouncy gaits, and the many nuances of human biomechanics.

"I'm sitting with the Dunkin' Donuts chew spit guys," Marcus says, "and this guy runs through the drill, he hauled ass, but he's pretty lumbering, not a great mover. But he smacks the bag, at the end, mega-hard. He hits it perfect—it made like a gunshot sound. The scouts all go crazy and scribble down his number."

Marcus looked right, looked left, and noticed that the brain trust of the NFL didn't care much about posterior tilt, but loved a gunshot sound. And he thought to himself, "OK, I am out."

# 4

OLYMPIC MOVEMENT

*Nature is pleased with simplicity.*

—Isaac Newton

STEIN METZGER AND KEVIN WONG finished the 2002 pro beach volleyball season as the sixth-ranked team. But the 2003 season would determine which two teams would represent the United States at the Athens Olympics. So, they'd have to step it up.

After years of sampling private trainers, high-profile coaches, and everything in between, Metzger felt he had finally perfected his exercise routine in the UCLA gym and was coming into the 2003 season strong. The six-foot, seven-inch Wong had been Metzger's teammate since the Punahou School back in Hawaii, and they had roomed together in college. Beach volleyball duos ideally combine a smaller, nimble defender, like the six-foot-three Metzger, who ranges all over the sand to keep the ball in play, and a blocker who patrols the airspace above the net. In August, halfway through the season, Wong and Metzger were in the hunt for the Olympics after a respectable second-place finish at a tournament in Austria.

Metzger flew to the next tournament in Prague; his phone rang as he got off the plane. It was his girlfriend, who read aloud from the *LA Times*: Wong had dumped Metzger. In the story, Wong said things like "it's a lot easier to get to the top of the mountain with someone who has been there." Wong would pair with Eric Fonoimoana, winner of Olympic gold in Sydney in 2000.

The *Times* called Wong and Fonoimoana "a new potential powerhouse" that "should start winning right away." Beach volleyball would be live on national TV for the first time, starting the following weekend. The article included a huge full-color photo of Fonoimoana next to bubbly talk from the tour's commissioner saying the sport needed "to create heroes for our fans." Maybe Metzger didn't count as a hero?

Metzger had prepared his whole life for this season, but he had not prepared to find a new teammate. One obvious candidate: Fonoimoana's just-dumped partner, Dax Holdren. The reason Holdren was available, according to Fonoimoana, was because Holdren was "just not healing enough to compete against the big guys." Holdren's knee was not the only concern: he was also, like Metzger, a six-foot, three-inch defender.

Metzger and Holdren got together, called themselves the Dumpees, and took stock. They had no blocker, two defenders, zero Olympic qualifying points, and little time. A doctor had told Holdren his days as an elite player might be over, but he was working out with a new guy in Santa Barbara. He urged his new partner to make the two-hour drive north to join them.

On paper, Marcus was broke. "I came out of medical school," Marcus says, "with $360,000 debt." The jobs he could get in professional sports, working for a team, paid fifty grand or so. He owed around forty grand a year on his college loans.

So, he decided to borrow more money. "I remember the rationalization," Marcus says. "I could go work these crappy jobs right now, and make very little, and pay for this stuff. Or I could go do what I want to do and hopefully get paid quite a bit more."

Marcus's big mission had, for the moment, a tiny footprint. Mike Swan, once Marcus's triathlon training partner and one of the best bike racers and triathletes at UCSB, had a physical therapy clinic in Santa Barbara. He agreed to rent Marcus a patch of floor in the back for Marcus's Peak Performance Project, or P3. ("How big," I asked, "was your patch of floor?" Marcus paused our hike and pointed to a fallen tree and said, "It was from *there*..."—he moved his pointer along to a twiggy branch—"...to *there*." It looked like about twenty feet.)

Sometimes he'd take clients to other places to work out. One day in the gym at UCSB, with a tennis player client in tow, Marcus stopped walking when he saw a blond woman on the stairmaster. Nadine was freshly arrived from Germany, had just joined the gym, and thought—as a child of the Alps, used to mountain climbing—she would try out this odd-looking American contraption.

But when she got on it, the machine was incredibly stiff. Nadine's English is perfect, but it's a second language, accented, and occasionally charmingly inventive. This machine, she said, was "clompy," just slapping her feet and jamming the stairs down.

"Do you need help?" was the first thing Marcus ever said to the love of his life.

She said she was fine, thanks. She was getting off that machine anyway, because she didn't like it.

Marcus nodded. "I'm not a huge fan of that machine, either," he said "but it works a lot better if you plug it in." Nadine laughs telling the story now.

A few days later, Marcus was out with friends, a married couple, and *there was Nadine.* By some miracle, Marcus's American friends had met in the tiny town in the Bavarian Alps where Nadine had grown up. The table filled with an Alpine glow. Soon Nadine was dating a Harvard doctor with huge plans and a tiny gym.

Marcus biked on Gibraltar Road above Santa Barbara—several thousand feet of breathtaking ocean views, sagebrush, and ten-foot-tall white-flowering chaparral yucca. One day in the early 2000s, Marcus walked up the driveway of the house with the best view and said hi. The couple who lived in the multimillion-dollar Spanish-style house said they'd always kept their guest house empty so that friends could stop by, but why not let Marcus and Nadine rent it?

They spent their weekends above the tree line. On weekdays, Marcus rolled out what he calls a "plyometric runway" in the back of Swan's physical therapy place, and built a practice training a mishmash of high school and college athletes—until Dax Holdren arrived as P3's first professional athlete.

Marcus remembers that Holdren had a "significant lesion" on his femur and arrived literally in tears, wondering if his career might be over. Something about what happened next gripped Holdren in a way that caused him to call Metzger again and again. "Just come up once and see what this guy is doing."

Metzger says he "didn't feel like he needed to drive all the way to Santa Barbara to work out." On the other hand, Metzger wasn't married, didn't have kids, and lived in a VW bus. He drove up in the name of being a good teammate.

P3 was way more than two hours, it seemed, from UCLA. Metzger says he walked past the physical therapy tables feeling "completely skeptical. Like, 'What is going on here? This is Mickey Mouse stuff.' "

Marcus got Metzger talking—about his body and his goals, then directed Metzger to move back and forth across the world's smallest gym.

"I just felt instantly," Metzger says, "like I'm totally doing the right things right now."

Metzger arrived with little faith and impeccable sensors. When things changed in his back or hips or neck, he had questions. Marcus had plain-English answers. "Your body moves *this way*, because of *that*," remembers Metzger. "We want to teach your body to do *this*."

P3's sole motion-capture tool was Marcus's two eyes. "Athletes would literally walk in the door," Metzger says, "and he could tell what that athlete needed, in terms of the holes . . . what they needed to plug in their system to be a complete athlete. That's crazy."

This kind of talk unsettles Marcus. He wants to drag the field away from snake oil and faith healers, and toward the world of evidence, the scientific method, and force plates. At the same time, he's the son of a sculptor who measures things in log lengths.

Marcus wanted Metzger to master the squishy-sounding "quality of movement." To Marcus, merely strengthening this or that muscle makes you a great athlete about as reliably as piano scales make you Thelonius

Monk. The key isn't hitting a note, it's linking together a thousand notes in thrilling ways.

A full orchestra has about ninety musicians, whose beauty emanates from the integrated efforts of the composer, the conductor, and the individual musicians' artistry, informed by practice. An athlete's body is like that orchestra, only instead of ninety musicians, it has six hundred muscles, conducted by the cerebellum, cortex, basal ganglia, and other brain parts integral to motor control.

Metzger's visits to Santa Barbara were, in Marcus's view, a chance for his muscle orchestra to learn new music. If the Dumpees were to dominate, they'd need Metzger spiking the living crap out of the ball. In explosive rotational forces like that, the body's soft tissues swing around the spine like Ginger Rogers grasping a pillar. Anchoring one end unleashes the other to fly. Many of the most fun movements in sports—hitting a baseball, throwing a javelin, spiking a volleyball—count on that kind of rotation. But Metzger had slumping posture that messed all that up.

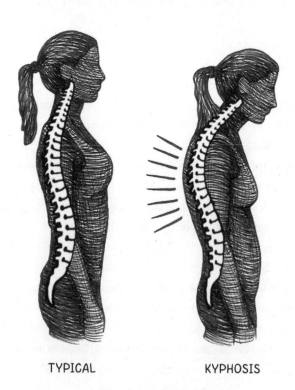

TYPICAL                    KYPHOSIS

No one stands perfectly straight. The upper part of even the healthiest back angles forward a little. Angle a bit more and my grandma might poke you in the kidneys at dinner. Tip forward more than forty-five degrees and it's diagnosable kyphosis. Long seen as the result of osteoporosis or violent accidents, nowadays kyphosis is often associated with using keyboards.

Marcus uses the phrase *computer back* more often than he uses computers. He calls it the most pervasive biomechanical issue among "normies." No doubt you see curved spines and hanging heads tilting at laptops or phones in your home and workplace. I notice it in NBA locker rooms and among professional athletes waiting for Ubers.

Kyphosis causes practical issues. "You can't reach as high," says Marcus. There aren't a ton of jobs where that matters, but volleyball player is one. "And you're not as strong. The vertebrae aren't stacked on top of each other with optimal confirmation and support from the paraspinal muscles. They're only partially stacked. It's not letting bones support each other if they're overlapping by 30 percent over the edge, one after another. It's not a strong structure." Marcus says people with kyphosis are often weak doing things above their heads. "They tend to get strains of the paraspinal muscles, and they have the head trying to shear down."

You can get away with kyphosis in straight-line sports like distance running. (Watch Olympic medal–winning mid-distance runner Timothy Cheruiyot.) But it inhibits explosive lateral movement, which is key for volleyball players, and which requires nervous system coordination akin to a conductor drifting an arm to summon the violins, a moment later the violas, and then finally the cellos.

Doctors treat kyphosis with surgery, body braces, special beds, or the "Schroth Method," which involves putting a pillow under the back and stretching the shoulders back toward the floor.

Marcus has a different approach: he rehearses the orchestra. "There really is a fix," Marcus says, calling kyphosis "very changeable. We have a bunch of therapies we throw at these, and it's not just telling people to bring their shoulders back. My one go-to is the snatch squat press."

SNATCH SQUAT PRESS

This is a move where you put a barbell behind your head, sit into a squat, and then remain squatted as you push the bar up with your arms.

For most people, squatting with a bar across their shoulders and then pushing up produces conflicting emotions. Obviously, they feel weight, but also alarm and confusion from the base of the neck to the bottom of the ribs.

If you have healthy posture, there is a reasonable amount of spinal support for your pressing arms, in the stacked vertebrae and the tissues that surround them. If you're kyphotic, you'll feel their absence. "Moving a long lever in a way that is not supported, it's really a muscular adaptation," says Marcus. "It's also putting pressure on ligaments and all the connective tissue. The entire system is being challenged." Sensing weakness in an unsupported movement, the nervous system tends to beef up the soft tissue along the spine, which pulls the vertebrae into a tidier stack.

"For people with significant thoracic curve," Marcus says, the snatch squat press is "super, super hard." Many start with a broom in place of the bar, say, or heels elevated on small blocks to reduce the angles involved.

Marcus says the snatch squat press is "basically forcing all of the body's wisdom to be applied to solve this problem."

The mistake, though, is to take that first-rep disorientation as a reason to stop. A lot of exercises challenge muscles. It feels weirder to challenge well-worn movement patterns.

Paired with foam rolling and strength exercises, Marcus says the snatch squat press produces results: "I've seen this fixed in so many people. Not just nineteen-year-olds. If you push on it enough, it just changes. Not like a multiyear fix, either. It can change a lot in a matter of months."

On Marcus's rolled-out mat, they improved Metzger's posture, thoracic movement, and kill speed. Metzger's VW bus came to feel at home round-tripping up the 101. Twice a week he'd camp somewhere beautiful, get up early to horse around in the ocean, show up a little salty and sandy at P3, and then refine the orchestral movement of his volleyball muscles.

One day, Metzger rented a boat to sail to the Channel Islands and invited Marcus. Metzger likes things pretty buttoned up, and he had plenty to worry about for this trip: the new boat, the wind, the fact he hadn't been sailing in a while, and the famously tricky navigation into the Catalina Island cove.

"The next thing you know, the sun's going down and we're only halfway there, because we were like, 'We need to sail, we're not motoring.'" They both remember Marcus, glass of wine in hand, steering the boat with his foot, blasting the Boozoo Bajou song "Night over Manaus." It's a drum circle of a song with a *slap-slap-slap* that could be waves on a hull. There basically aren't words, other than a shamanic chant. "We may," Marcus says, "have smoked a joint." They arrived in darkness but had a blast.

In some ways, Metzger's kyphosis was the Dumpees' easiest problem. They had to save Holdren's career by teaching him how to move in femur-friendly ways, and oh yeah, as they were both natural defenders, someone would have to learn to play out of position, as a blocker up against the tall opponents at the net.

Metzger drew the short straw and would learn to get long. For this project, Marcus assigned two-legged Olympic lifts, to get a very solid base to jump from the standing position, which is what happens at the net—and which is a big change from leaping laterally off one foot, digging balls out of the sand.

And every weekend they were off to battle for Olympic qualifying points. "When I met him," Metzger says of Marcus, "this was a phase in my career when I was doing pretty well. I was getting, like, some fifths and would maybe squeak out a win now and then." With Kevin Wong, Metzger won the 2002 Espinho Open in Portugal. "I started training with [Marcus] and all of a sudden I was in final fours consistently, every weekend." He says that "the first eighteen months I spent with Marcus . . . translated to real, I mean *real* money, for me for the first time."

In October, the Dumpees arrived at the world championships on the white sands of Copacabana Beach. The stands were a rolling dance party, the weather so hot the organizers soaked the crowd with sprinklers. Holdren and Metzger had been paired for less than two months, and arrived as the fifteenth-seeded team. Dain Blanton and Jeff Nygaard would likely take one American Olympic spot—the Dumpees and dumpers would fight for the other.

The Dumpees' taller opponents forced the action to the space above the net. But the Dumpees were crafty as hell, and used all manner of doinks and backhands to squirrel the ball over and around the long arms of natural blockers. They racked up a fair number of points in the early rounds, and found themselves on the main court in the semifinals, against the beloved, second-ranked home team of Márcio Araújo and Benjamin Insfran.

Almost everything Metzger worked on at P3 manifested in crunch time. Metzger leapt off two legs for a ball at the net, and encountered the bigger, longer, vastly more explosive Insfran. The Eurovision broadcaster, wittingly or not, touted a victory over kyphosis: "Great swing from Stein Metzger! Arched his back and really unleashed on that one!"

The game's final point was almost in slow motion: Metzger crouched before the net, hips back and low, as if he were preparing to do an Olympic lift on the rolled-out mat in P3's corner of the physical therapy studio. As Insfran skied to spike, Metzger leapt, too—up and, wisely, a tiny bit left, where he sealed the victory with a block. An undersized player led the match in blocks. The Dumpees finished second in the world championships. The points windfall was more than enough to get them to the Olympics.

Fonoimoana and Wong—the dumpers—finished in an eight-way tie for ninth.

At the Athens Olympics, the Dumpees finished tied for fifth overall, by far the best American team. Then they both found traditional blocking partners. Metzger worked with Marcus to transition his body back to lateral explosiveness. That meant a lot of single-leg movements, especially explosive sideways leaps known as *lateral plyometrics*. The classic hang clean lift was not new to Metzger, but he gets a religious tenor in his voice when he describes how serious he and Marcus came to be about that barbell maneuver. It starts with knees slightly bent and the weighted bar hanging across the front of the hips, arms straight. And then there's a timed explosion. Almost every big group of muscles in the legs (hamstrings, glutes, calves, quads) pushes up, as if jumping, as upper-body muscles (the shrugging shoulders, and pulling trapezius and biceps) work to yank the bar up. It ends by catching the bar across the front of the shoulders, lowering into a deep squat, and standing again. Metzger added weight and repetitions, and felt his body becoming twitchier, more electrified.

For two straight years, with two different partners, Metzger was on the top-ranked team in the US, with Holdren ranked second. "It felt like I went from trying to perform on a race course in a VW bus to a Porsche," says Metzger. All his career, he'd been able to see, in real time, balls that were just out of reach. "And all of a sudden, I could get there."

# 5

## THE DUMBEST SPORT

*If you challenge conventional wisdom,*
*you will find ways to do things much*
*better than they are currently done.*

—Bill James

SIX-FOOT, FOUR-INCH ALL-AMERICAN linebacker and tight end Steve Zabel often made the newspaper for his "smashmouth" play. But on a windy November weekend in 1969, Zabel made it into print for crying. His Oklahoma Sooners had lost—badly, unexpectedly—to Nebraska.

One sportswriter called the Cornhuskers' offensive line "Nebraska's muscle curtain." At one point, Nebraska rattled off five straight touchdowns and a field goal. Quarterback Van Brownson—not known to be especially nimble—ran for eighty-two yards. In newsreel footage, the Cornhuskers appeared immune to tackles.

Two things died that day: Oklahoma's hopes for a spot in the Sun Bowl, and the idea that football players shouldn't lift weights. Conventional wisdom at the time held that pushing around heavy iron would make players "muscle-bound"—immobile as Clydesdales. The few team-sport athletes who lifted were referred to as "barbell men." (Sergio Oliva won the 1968 Mr. Olympia bodybuilding competition, and its thousand-dollar prize, because *nobody else competed*.)

The Nebraska team was a few months into a groundbreaking weight training program led by an asthmatic pole vaulter named Boyd Epley. Epley's family was from Nebraska, but he had spent most of his childhood in Arizona, where his family moved in the hopes the desert air would help his breathing. Epley first lifted weights as a ten-year-old in his buddy Danny's garage. A couple of years later, his dad bought him a weight kit that came with a brochure that taught fundamentals like dead lifts, bench presses, and squats. Epley put on twenty pounds over the summer before his senior year and became an elite high school linebacker and record-setting pole vaulter for the junior college national champions, Phoenix College. Epley transferred to Nebraska and set pole vaulting records indoors and out.

Then Epley hurt his back, stopped competing, and innovated his own recovery as the most devoted user of Nebraska's cramped and rudimentary weight room. After a few injured football players grew curious about Epley's methods, he got a call from one of the football team's assistant coaches. Epley thought he might be in trouble for teaching football players to lift. Instead, he got two dollars an hour, two days a week, to overhaul football training—and a warning that if any of the players got slower, he'd be fired.

At first, they barely had any equipment. Epley remembers running players through snatches, cleans, curls, and presses with a single 47.3-pound bar weighted by paint cans full of dried concrete. To earn trust, Epley resolved to measure everything from maximum bench press to vertical jump. To prove players weren't getting slower, he raced players against the clock in the forty-meter sprint constantly, and showed the times to the coaches.

The Huskers didn't slow down as they became absolute beefcakes. *Minneapolis Star Tribune* columnist Joe Soucheray visited Nebraska's gym and reported seeing any number of football players, "none of whom had necks and any one of whom could play the villain in the next James Bond movie."

One of Epley's early battles was to eliminate what had been the core of the team's training before his arrival: five-mile runs. Football, Epley says he explained in increasingly exasperated tones over his first few years, has no aerobic component at all. A play lasts about five seconds, typically followed, Epley explained, by a fifty-second rest. Fifty seconds means a well-trained athlete can start the next play 100 percent fresh. The team stopped running so much, and kept winning.

After the win at Oklahoma, the Huskers beat Georgia in the Sun Bowl, followed by national championships in 1970 and 1971, and three more titles after that. Epley became the godfather of American strength coaches.

Epley assigned himself three years to collect and assess lessons from the bleeding edge of weight lifting. Mostly, he learned by visiting competitions, entering, and very often winning. Epley was crowned Mr. Nebraska three years running. This made Epley an expert in the slightly different fields of powerlifting (lifting the heaviest possible weights), bodybuilding (showing off muscles), and Olympic lifting (honing a tightly defined set of explosive moves). Nebraska's football players squatted, benched, and deadlifted like powerlifters, but also grew explosive with Olympic moves like the power clean and push jerk.

Epley inched closer to the center of the sports world. He designed exercise equipment and a Reebok shoe. He formulated supplements. He wrote books. He created an athleticism index. He designed a massive new training facility for Nebraska, and consulted on another, for Miami.

In 1978 Epley founded what is now known as the National Strength and Conditioning Association. An entire industry blossomed in his wake. Evangelists of weight lifting can now be found wherever there are elite athletes, even in tennis, golf, ballet, Formula One, and video gaming.

Epley's hot knife of insight broke up ice floes of groupthink. Epley put to rest the idea that fast and mobile athletes couldn't benefit from strength. This could have resulted in deeper curiosity: What other assumptions about the human body are wrong? What else did science have to tell us?

Instead, to Marcus's eyes, the berg of groupthink refroze in a slightly new shape—more or less around insight from the brochure that came with Epley's weight set. Today the organization Epley formed, the NSCA, says it has more than 60,000 "members and certified professionals." In Marcus's experience, they're as ardent about big muscles as football coaches once were about long runs.

Meanwhile, as Marcus was having force plates installed in P3's new, expanded location, clues emerged that traditional strength training only captured part of the story. Baseball pitchers who can throw the ball ninety-five miles an hour fell into two groups. There were enormous, strong men who would ace NSCA tests like the bench press, but there was another class of player, too. This second group tended to be much smaller than average players, and unremarkable in strength tests. But, fascinatingly, they scored exceptionally well on the lateral skater test. Their super-skill appeared to be carrying force from the ground, up through the legs, through the many planes of a twisting torso, and down into the arm and, ultimately, the ball. They had something cooking in the nervous system. And, interestingly, those players had almost all grown up pitching.

Which dovetailed with another baseball mystery: Why was it that every elite player in this sport began playing before the age of twelve? It's not a simple case of practice making perfect in sports. Many NBA MVPs—Tim Duncan, Joel Embiid, Steve Nash, Hakeem Olajuwon—pivoted to basketball only after puberty. But professional baseball players all grow up playing baseball. And some smaller athletes can throw the ball ninety-five miles an hour without being superstrong.

Marcus realized that both mysteries emanate from the same bedrock: baseball is a rotational sport. Rotating is a kind of athleticism barely considered by traditional strength training. "Traditionally, strength coaches train strength. Guys came in, they lifted weights, they left," Marcus says. At P3, though, they "were teaching movement."

In a perfect world, your brain would have automated competent rotational movement, just as simply and instinctively as we open doors, walk up stairs, or get out of bed.

But for many of us, instead, we have to think about how we move through complex things. This dynamic is especially obvious to anyone in pain. "Part of it is broken," Marcus says, remembering how knee soreness caused him to struggle in his weekly yoga class. "You have to just piecemeal that whole thing together. It's so tedious. It takes so much brain function. And it's really like instead of one movement, this part, that part, that part."

The goal of a lot of P3 training is to make healthy movement automatic. "You just code for that rotational movement, and so you just reach into that movement vocabulary you have to grab that thing and you do it," he says.

P3 has dozens of rotational exercises. One favorite is to lie on your back, arms splayed, with a dumbbell anchoring each hand. Lift your two legs together, soles facing the ceiling. With shoulders flat to the ground and legs glued together, swing your feet to the left side and twist so that ten toes touch the floor, then lift them high again and do the same on the right. At P3 I watched torsos rotate many different ways in every warm-up and workout, every day. They frequently recommend a clever thoracic mobility move where you sit on the floor with knees bent in front of you, and lean back to press the middle of your back into a foam roller. Then let your knees fall together to the right, and then the left, three times each. Then move the roller up your back one vertebra and try again. It lets you choose which segment of your back to twist.

There's also rotation with force, from throwing medicine balls sideways to pulling cables with heavy resistance. These are activities you may have encountered at a CrossFit class. But the chop bar march is not something you're likely to find at your local gym. With a back to the Keiser cable machine, an athlete holds a bar attached to the cable in two hands, off to the right. Then, in one smooth movement, she pivots her torso so she's facing forward, hands on the bar out in front like she's water-skiing. As she rotates, she also lifts her left knee to hip height. It ends up being a rotational core muscle exercise *and* a balance drill— entirely typical of P3.

This is the kind of stuff that was on Marcus's mind when he got an invitation to speak at baseball's winter meetings to the Professional Baseball Strength and Conditioning Coaches Society—a group of NSCA-certified professionals who had banded together to professionalize strength training in baseball.

Marcus accepted, flew to Indianapolis, and stood at a lectern, thinking this room full of men who revered Boyd Epley might also love the cool science Marcus had to share. An early concern: "Looking out, I saw big necks, square shoulders, bald heads," says Marcus. "All guys who have more testosterone than is good for them." The finest players in the world's most rotational sport were being instructed in movement by men who could not turn their necks.

The NSCA's fixation on big muscles dovetails with a fuzzy position on anabolic steroids and performance-enhancing drugs. The USA Strength and Conditioning Coaches Hall of Fame celebrates, for instance, pioneering coach Alvin Roy as "the first coach to administer (legal and still not stigmatized) anabolic steroids to players." Louis Riecke's biography describes his advanced workouts as a young man, but adds "by this time Riecke realized his spectacular gains resulted more from the daily anabolic."

But getting stronger, Marcus says, can make you *worse* at sports, because tasks like hitting and throwing are multi-segmental. There's pushing from the ground, from the hips, the torso, shoulders, and arms. "You can get really strong in each segment and still not create a lot of rotational power," he says. "And now you got all this weight to carry around, which makes it hard to change direction. Beats on joints more. It's easy to do a whole lot of work on your shoulders and your deltoids and lose mobility in your shoulder and not be able to stick your arm straight up over your head."

Pushing big stacks of weights has another issue. "On the nervous system side," Marcus says, "you train to do things slowly, and your body starts thinking you need to do things slowly. Nothing about that is gonna make you faster, more explosive."

As Marcus took the lectern, he was talking to professionals caught in the rising tide of baseball's steroid scandal. Baseball's strength coaches had leveraged the early days of the scandal to get personal trainers banned from clubhouses. But in the years that followed, the scandal mushroomed.

The training group's first president, Fernando Montes, had been the Cleveland Indians' strength coach when Canadian authorities found drugs on a team plane. Montes also worked for the Texas Rangers when their slugger Alex Rodriguez was using PEDs.

Marcus spoke three years after the book *Game of Shadows* shocked baseball by reporting that the best player in baseball, Barry Bonds, had been inspired to dope by watching other beloved stars Sammy Sosa and Mark McGwire. He spoke two years after the league's Mitchell Report exposed the breadth and depth of the problem, and shortly before McGwire—who had trained with people in Marcus's audience—confessed to doping.

Meanwhile, at these very winter meetings, word spread that Brad Pitt had signed on to star in the movie of Michael Lewis's book *Moneyball*. That meant there'd be a major motion picture based on a best seller that cataloged baseball's tired and outdated methods. The billionaire owner of the Boston Red Sox, John Henry, was at the meetings, making noise about teams being backward.

Marcus brought an antsy audience a message that their ways were outdated. "I'm coming in with a lot of conviction," Marcus remembers. He opened by asking for a show of hands: Who had their pitchers go on "flush runs"? Long runs to flush lactic acid from pitchers' bodies were an entrenched tradition in baseball. Many hands went up.

Marcus said they should stop at once because pitchers don't generate much lactic acid. Channeling Epley, who told the Nebraska coaches that five-mile runs had no place in football, Marcus explained that a pitch is basically a one-meter sprint. Bodies don't generate much lactic acid at such distances. Anyway, lactic acid clears quickly, with or without a run. By the time these legends of pitching were lacing up their running shoes the morning after a big game, there was no lactic acid left to flush.

But it was worse than that. After a certain level of fitness, maximizing muscles for endurance comes at the cost of explosion. "It's not just not helpful," Marcus says, for one-meter sprinters to run five miles, "it's interruptive." That term hangs in the air before he adds: "That means it's harmful." He finished his speech, suggesting that the work isn't to be in better overall shape. It's to tune your body to its precise task. None of the coaches seemed to care.

Marcus says he genuinely thought these coaches would fist-bump him for helping to improve the performance of ace pitchers like Tim

Lincecum, Cliff Lee, and CC Sabathia. Instead, he felt hostility. This was later confirmed when Marcus met a strength coach who had heard Marcus speak in Indianapolis. "He shook my hand and said, 'You might be fucking smart, but I fucking hate you,'" remembers Marcus, who could barely respond. "I was just too floored by it."

Marcus was out of sorts in his hotel room after the speech. "I felt kind of dirty," Marcus remembers. "Cultural whiplash, coming out of five years at Harvard." Marcus's body had an ironic idea: a flush run. But in his carry-on the Californian hadn't packed winter running gear, and that weekend, Indianapolis had pause-inducing fifty-mile-an-hour winds, temperatures below zero, and intermittent freezing rain.

*What the hell*, thought Marcus. "In the middle of the run," Marcus remembers thinking, "I was astonished at how much these guys didn't like me. My first instinct was to get away."

The frigid air cooled him. Maybe the wolf pack's reaction wasn't personal. "People are feeling insecure about this earthquake that just went through their world," he says. "The more right you are, the bigger the earthquake."

And there was another reason the strength coaches might have been rattled: they thought Marcus, or someone like him, might take their jobs. And in fact, Marcus was sitting on MLB job offers. Confusing feelings coalesced on the run. "I got it clear as day," Marcus says. "If you're going to practice your craft, you stand up. If they throw more tomatoes, you have to yell a little louder." He would sign up with a team. "We gotta go show these guys. Be a beacon."

In 2010, Marcus became the Seattle Mariners' director of sports science and performance. In 141 years of major-league baseball, no one had held a title like that. Marcus negotiated the position so that he could do it while also running P3. He'd make regular visits to the Mariners while his handpicked staffers spent the season with the players.

The first thing Marcus did was to remove almost all of the equipment from the Mariners gym, giving players room to jump, run, and lunge. "They had three different calf raise machines," he notes before almost yelling, "SPORT IS MOVEMENT! And you couldn't take a long step in that building."

The Mariners would move, especially in ways that reduced players' *stretch-shortening cycle*. Athletic movement often involves changing directions. Wind the bat back and then explode forward into a home run swing. An outfielder brings the ball back before hurling it forward. Many of the fastest people—NBA players starting their drives to the rim, MLB players stealing second—make their first step backward. It's often called a "false step," and coaches have long trained it away by putting an empty soda can behind the athlete's foot—it clatters away when the athlete's foot involuntarily moves backward before going forward. The surprise has come from a flood of biomechanical research that shows the false step gets most athletes to their destination *faster*, because of the stretch-shortening cycle in that planted leg.

Research from the earliest days of sports science in the Soviet Union had demonstrated that you could generate more power by training the two movements, in opposite directions, into one snappy event. A classic example: most people can jump higher by running into a jump, because they can translate a little downward force of landing into the upward movement of jumping. A whole school of training developed to make those movements faster and more efficient. The Soviets called it the shock method; in the US we call them plyometrics, and Marcus loves them to this day.

As plyometrics caught on in the United States, the NSCA published guidelines saying athletes should only perform plyometrics once they are strong enough to squat one and a half times their body weight. "That was the standard that was held out for a long time," Marcus says, "and kids are doing plyometrics all the time. Jumping out of trees! They're playing tag! They're skipping! They can't squat one and a half times their body weight, generally, or maybe ever. And they're all doing plyometrics. So, that's not the problem. The problem is just mechanics." In practical terms, it meant Marcus watching closely, and coaching baseball players to be bouncier, quicker, and more explosive.

Marcus ruffled feathers by telling the Seattle beat writers that some of the Mariners "don't seem like they should make a college sports team, in terms of athleticism." The beauty of the wide-open gym, Marcus said, was that unathletic players would come in and do big, explosive movements, and "there's nowhere for them to hide their lack of athleticism. It just stands out like crazy."

When spring training kicked off in Peoria, Arizona, Marcus got a chance to meet his new coworkers all together for the first time. Looking around the striped green infield, he saw "all these middle-aged guys with Dunkin' Donuts coffee cups, which they were using to spit chew into. They wanted nothing to do with me. I could see it was going to be a struggle."

One of his new bosses cautioned Marcus to go slow—he said it would take a decade to affect real cultural change. "I didn't have a decade," Marcus says. "We're not curing polio here. If you can't make it measurably better, if you can't really move the needle—what's the point? . . . The normal routines were violating laws of biomechanics and needed to change."

Marcus escaped to the cheap seats to make two phone calls. "I remember crawling up into the top of the stadium and calling my dad. I called a Catholic priest I've been close to for a long time. I just wasn't sure I could do it."

Ultimately, Marcus decided to stay and began building his staff. One of his hires in Seattle was a scientist named Charles Kenyon. The bulk of Kenyon's research had been in phylogenetics, digging deep into the relationship between orchid species. "Great," Marcus remembers thinking, "we got the orchid guy."

But Charles had an impeccable scientific mind, experience at two elite performance facilities, and plans to go to medical school.

Marcus asked Charles, and former pro player Danny Garcia—who had trained at P3—to be the faces of P3 in Seattle. Charles had been working with baseball players at P3 and loved it, and was from Seattle, so he deferred medical school and moved.

Marcus had negotiated dominion over the minor leaguers in the Mariners organization but had less influence over the major-league players. (Mariners star Ken Griffey Jr. responded to Marcus's arrival by saying he would keep working out the way he had before. Griffey missed 129 games that season.) For those on his program, Marcus banned flush runs and added a ton of rotational movements, plyometrics, and explosive lateral movements. He also integrated overhead weights, which baseball had long treated with superstition, and the Keiser cable machine, which almost no one in baseball had seen before. Soon, players were holding the yard-long handle and chopping away to gain rotational quickness through the strike zone.

Veteran outfielder Eric Byrnes told reporters he loved feeling explosive. Minor-league third baseman Kyle Seager embraced everything about P3 and became a major-league All-Star.

Charles is a doctor now, a physiatrist at the University of Washington, and says he's proud of P3's work with the Mariners, even if they "kind of rubbed against the culture and the tradition." Marcus wanted players to reduce their reliance on traditional weight lifting and learn how to *move*. Charles says the staff learned something, too: that the science meant nothing unless "you can have that conversation with the athlete" and "translate all the findings in a way that he or she cares about." If you can "convince them that's the first reason you're there, and they feel supported, and that you're connecting with them deeper than just the athlete level," Charles says, "I think that's where having that data becomes really powerful."

The immovable rock was the strength coach, who, not surprisingly, resisted Marcus's efforts.

The conflict came to a head over the high-salaried pitcher Érik Bédard. "Left-handed pitcher, paid a fair amount," remembers Marcus. "And he's coming off elbow and two shoulder surgeries. He was so promising on the mound, but had missed two or three seasons. I saw him getting ready for his third bench press workout of the same week. He was super-overdeveloped in his chest and biceps with his knucklehead workout program. His poor posture was putting load on his shoulder and elbow—this wouldn't help."

Marcus could see that Bédard might easily get hurt again, and keeping the pitcher healthy was his job. "I laid it out to them. What I got back was: 'I don't like people who try to figure things out.' That came from the player, but standing next to him was the strength coach who looked very satisfied."

Whether or not the strength coach had directed Bédard to say this, it was very clear that the strength coach wasn't going to let Marcus replace or reduce the strength workouts. The ideal solution for Marcus would have been to remove the strength coach, but the GM made clear to Marcus that this wasn't an option. Marcus finished his three-year contract and says that "baseball is definitely the dumbest sport."

# 6

......................

## LAND

......................

*She could not grasp exactly what to do next,*
*but she kept moving as if her life depended*
*on it, which in some ways, it did.*

—James McBride

WHEN PROFESSIONAL BEACH volleyball player Katie Spieler was tiny, her family would drop her into a Boston Whaler and sputter out of Santa Barbara Harbor. Their destination was often the Channel Islands, five peaks that poke out of the Pacific, *Lord of the Rings* dramatic, dotted with private coves and beaches where a young family can cut the tethers of civilization.

Katie spent whole afternoons there, snorkeling among the sea cucumbers and Garibaldi fish. She leapt off boulders, surfed waves, and climbed sheer faces. She darted about like a bottlenose. Nothing felt impossible.

At home in Santa Barbara, Katie and her sister Cara spent their days in the sun—in the San Marcos Foothills and on East Beach. The only problem for Katie was that her two favorite things—beach volleyball tournaments and the Stud Ironman beach lifeguard competition—fell on the same Tuesday. Katie made it work. "I would go play the beach volleyball tournament, and I would, like, win that, and then I would go and race over down the beach to junior lifeguards." Katie's friend Chrissy, a counselor, would run a second session just for Katie. "Eight times you do a swim, out to a buoy," Katie says, "run to the next lifeguard tower, swim the buoy, and do the same thing all the way to the pier and back from East Beach." Chrissy and Katie are still friends.

"That was, like, my favorite day of the year," Katie says.

Legends of volleyball like Kathy Gregory and Karch Kiraly came through East Beach. Little Katie noticed how tall the best players were, and hoped she'd grow. The women's net is 7.35 feet in the air; the best hitters play well above that. Sometimes Katie teases her mom for "marrying a short guy." Instead of blocking balls at the net, Katie became a defender. In a typical action photo, Katie is sandy and horizontal. Serving and placement skills made her a star at the powerhouse University of Hawaii. When she graduated in 2016, she became an inspiration to normal-sized players everywhere by joining the AVP Pro Tour at five-foot-five—a full foot shorter than some competitors.

Pro beach volleyball doesn't generally offer glamorous salaries. To make ends meet, Katie and her Hawaii teammate Dana Kabashima started a volleyball academy on East Beach, home of Katie's one-girl Stud Ironman. It foundered in its second season. "We always laugh about it," Katie says, "but we had *almost* one student." But while it was in operation, Katie and Dana printed up fliers and looked for athletes, eventually finding themselves in front of P3. P3 had graduated, in 2009, from the physical therapist's office to a former nightclub space in the Funk Zone, a revitalized industrial neighborhood just off the beach. Katie knew that professional athletes, including NBA and MLB players, worked out there. And she had met a couple of P3 trainers on the volleyball court. They seemed serious about their jobs—and pretty good at volleyball, too.

But the place had neither a sun-splashed reception area nor any kind of front desk. P3 didn't even have a sign—just a black logo on white walls and the muted pulse of hip-hop from within.

Entering felt like slipping into church with the service underway. As their eyes adjusted, Katie and Dana heard hollers over the music, and saw they were standing at the start line of a track that sloped up to the back wall. To their right were nine scuffed wooden steps and a raised area with an L-shaped black faux leather couch. Most of the light came from a row of high windows down the right wall.

In front of everything, down three steps from the couches, running the width of the entire building, lay the star of the show: a fifty-by-sixty-foot open black floor with the high ceilings and airy feeling of a dance studio—but dotted with cable machines, squat racks, plyometric boxes, and other strange gym equipment.

A workout was ending. A trainer in a black T-shirt chatted with three long-limbed, lean-muscled people in the sweaty afterglow of hard work.

No one seemed to think it was weird that two stray volleyballers had wandered in. Katie recognized the trainer from the beach and said hello. Sam Brown was easy to talk to and curious about volleyball. He walked them around and pointed out the force plates, which looked like the checked-bag scales at the airport. Eventually, Katie and Dana handed over a fistful of fliers, nice and easy.

On the way out, Dana turned to Katie and asked, "Why don't you train here?"

LATERAL MEDICINE BALL TOSS

*I do my own thing*, Katie remembers thinking impulsively, reflecting on her intense bike rides, swims, and hill sprints. *I'm a very self-made athlete.* To any normal observer, she's a portrait of lean muscle. But Katie has always felt that her advantages were mostly in her head and heart: a willingness to learn, to work, and to adapt.

On the pro tour, though, even the tall players had Katie's grit and adaptability. Maybe Dana was right.

Katie arrived for her first session to booming music and a workout in full swing. A lot of gyms write one workout on a whiteboard for every athlete to follow. P3 doesn't have a whiteboard. Here, it seemed no two people followed the same program.

A hundred times at other gyms she'd seen someone pick up a heavy medicine ball, usually to toss it up high off the wall and then catch it again; a "wall ball." At P3, she watched a guy—was he a major-league baseball player?—who started with a heavy medicine ball in two hands on his chest, but then he whaled it *sideways*, grunting angrily, hard into the cinder-block wall. A regular wall ball is said to work quads, glutes, and a few other muscles. This . . . it was hard to even imagine the muscles in play.

The ball smacked the wall. A screen displayed the velocity: thirty miles an hour. "Awww." He was bummed.

Across the room, a coach whooped and hollered, cheering as a football player pounded out salsa-like rhythms on a plywood platform. A woman held a heavy plate as she dipped into a lunge—that's pretty normal—but she had the heel of her front foot hovering off a ledge, which was new to Katie.

Katie says she "honestly had no clue what they were doing in there," but understood, "Okay, this is more than just a gym."

Sam grabbed a clipboard and directed Katie to a physical therapy table. They went through her medical history, which didn't take long as she'd never really been hurt. What happened next was like a physical, but instead of listening to lungs with a stethoscope, he pushed her heel toward her butt, first on one side, then the other. She lay on her back, with one leg in the air and knee bent; Sam put a hand on her knee, to hold it in place, and then swung her foot like the arm of a clock. He wrote down everything: how far it could easily swing inward toward her midsection, and outward over the floor.

Movement to Katie had meant leaping from rocks or snorkeling or flying over the sand to dig out a hard-hit ball; it felt strange to reduce all that to Sam's clipboard.

Sam had Katie face a patch of wall near the bathroom door. She bent her leg to touch a kneecap to the white-painted wall. Easy. Now, he said, move your foot back, keep your heel locked to the floor, and touch your knee to the wall again. There were numbers marked on the floor, like a ruler. He found the furthest point where she could still touch, and wrote it down, then held a little gizmo to her shin to measure her shin angle, and wrote that down, too.

Then they got to hopping. She faced forward, with a small dowel on the floor parallel to the outside of her foot. The goal was to leap sideways over the dowel off two pressed-together feet (easy!) and back (great!) as many times as possible in eight seconds (mind-melting). She managed thirty-two hops in eight seconds.

"That was pretty intense," Katie remembers. The dowel test felt like reacting, on the beach, to a hard-hit volleyball; she felt she should be good at it. But later, she watched a video of some kid with bleached hair, black socks, and a white shirt with SOCCER across the front. He set a P3 record by hammering across that dowel *forty-three times* in eight seconds. She felt self-conscious about her performance. "All the other athletes are working out and then you're getting tested and then everyone's kind of watching you." She remembers thinking, *Okay, here's going to be not their top assessment!*

Sam nodded to the back of the room, where Katie met the people running the computers. With great care, the black-T-shirted P3 staffers stuck twenty-two reflective markers on Katie's feet, shins, knees, thighs, hips, and torso like she was going to be in a video game. Then she spent fifteen minutes doing whatever weird movements they told her to do, while technicians hovered over computer screens learning things about her.

Here the music seemed louder and the other athletes closer. Just behind Katie, in a giant floor-to-ceiling rig, a very strong man squatted with a bar across his back, weights on the bar, and an enormous heavy chain curling like a boa constrictor, up from the floor, across his shoulders, and back to the floor on the other side.

Much of the assessment was jumping. So much jumping! Jump as high as you can off the force plate however you want. Then off two legs, now off one. Step off an eighteen-inch red-metal plyometric box, land on two feet on the force plates, then jump as high as you can. Katie remembers thinking that it would be awesome if, after training here for a while, she could jump like Michael Jordan.

The feeling crept in that the stakes were high. Professional athletes flirt with professional mortality all the time. It struck Katie that the biomechanists in black T-shirts, looking at computer screens, might know more about her athletic future than she did.

Sam directed Katie to leap off one leg, then land on that same leg on the force plate. Okay fine, but the test wasn't to see how far or high she could jump. It was how quickly she could settle into stillness after landing.

Then, with her left foot on the force plate, Sam told Katie to *explode* hard and fast to her right. This "one-off skater" is part of every P3 assessment, and measures movement crucial to playing NBA defense, stealing a base, or Katie's calling card: reacting to hard-hit volleyballs.

Around then, Katie began to sweat. She suspected she wasn't getting an A. "My strength is figuring it out, learning something and learning it quick, and then getting good at it quickly," Katie says. "But with those things, it was kind of hard for me because I don't know what to do to actually make this better in the moment." No one studies for this pop quiz. One of the drills included a big wooden dowel—a giant's broomstick—across her shoulders. Then she rotated left and right as much as she easily could, measuring the rotational mobility of her upper spine. How do you ace that?

When it was finally done, they invited her to peek at the computer monitors. On video, Katie stepped off the box and landed on two feet. In another window on the computer screen, her animated skeleton moved like a dancing X-ray.

A third window contained a comedic opera of numbers—too many to make sense of, even with Katie's degree in accounting. How hard did she land on her right foot versus her left? How long was she on the ground before exploding upward? What speed did she explode out of the skater to the right? They could go all day.

The P3 staff did not say, "Soon you'll be jumping like MJ." Instead, there was the whiff of worry. These nice people, she felt, had bad news.

Katie was one of the smallest athletes P3 had ever assessed. But when she stepped off an eighteen-inch box, she landed with a peak force of 3,100 newtons. P3's head of biomechanics, Eric Leidersdorf, remembers it sounded "a little bit like, I don't know, a sack of bricks hitting the ground." It takes about twenty-five newtons to spike a volleyball, and double that to throw a baseball ninety miles an hour. A punch from a professional boxer packs about a thousand newtons.

Thirty-one hundred newtons isn't a number a human body can readily generate. It takes, say, a three-hundred-pound male bighorn sheep. "Mating competition involves two rams running toward one another," says the National Wildlife Federation website, "at speeds around 40 miles an hour and clashing their curled horns, which produces a sound that can be heard a mile away." That's two rams, each a hundred pounds heavier than Usain Bolt, and almost twice the speed. The bighorn sheep, the website says, connect head-on with a "maximum fighting force" of 3,400 newtons.

In other words: landing is hard. That was something P3 measured more than anyone. When P3 moved to the former nightclub, Marcus got a new loan to install side-by-side force plates in the floor, encased in tons of concrete. "It's like a scale," Marcus says, "but it's a really precise, complex scale that gives us information in all these dimensions—not just down." Marcus had never heard of another training facility using force plates, but he thought it would be enlightening. "If we understand how you push into the ground," says Marcus, "then it can teach us a lot about what's happening above the ground."

One thing they learned is that athletes push into the ground hard. Katie's alarming numbers demanded attention, but were hardly unseen on P3's force plates; the house record is 11,000 newtons.

Maybe you remember the egg drop competition, where a physics teacher handed out tape, straws, balloons, paper cups, and the like? The

assignment is to build a contraption to keep an egg from
breaking when tossed from
some window or ladder.

Katie's not an egg, and
an eighteen-inch plyometric
box isn't a second-story window. But an egg and a volleyball
player both free-fall with sufficient kinetic energy to shatter. High
school students cushion their egg's
landing with plastic shopping bags as
parachutes, balloons, sacks of popcorn,
rubber bands—you name it.

Humans in free fall have a different
set of tools: legs. To keep our heavy heads
and torsos from cracking, our legs evolved
over eons to bend in three places: the
ankle, knee, and hip. To a biomechanist,
each joint is a hinge, supported by strong
tissues like the Achilles, calves, quads, and
glutes. There are ways those tissues can work
beautifully together and absorb forces that would
otherwise be devastating.

Katie had grown up playing on soft sand, and landed with
her toes pointed downward. That's an absolute red flag to P3. The problem is that after the toes absorb the first little whiff of force, the middle
of the foot, which is wired to some of the strongest soft tissues in the
body, barely participates. Then the remainder comes crashing down on
the only part of the foot that's bony, hard, and noisy: the heel. P3's biggest impact numbers always come from heels.

Just as there are many ways to tape together drinking straws that
won't protect an egg, there are many dangerous ways to land. Imagine
someone painting an analog clock on the ground. Imagine that Katie
lands with her foot in the dead center. Where's the knee? What time

does it tell? Ideally, none: the knee would hover over the ball of the foot at the heart of the clock face while the weight of her torso falls onto her sinking hips. But in reality, those knees often lean and point to times like three o'clock or nine o'clock, each of which raises concerns.

A strong lower leg changes things a bit. An NFL wide receiver curling around a cornerback could plant his right foot in the middle of the clock and keep a healthy knee at nine o'clock. The foot can even tip a little, with the inside raising up. But there's a line the P3 biomechanists call the "asymptote of doom" after which the foot just gives up and rolls into a classic ankle sprain.

Issues can also arise before that point. Just landing with your knee a little bit toward three o'clock tests the slender little fifth metatarsal on the outside of the foot. When this breaks, they call it a Jones fracture, and it's one of the NBA's more common injuries, having afflicted Kevin Durant, Brook Lopez, CJ McCollum, and Rasheed Wallace. And so it goes around the clock. Put too much force on nine o'clock, and you'll start to hear people worrying about the anterior cruciate ligament on the inside of the knee.

Then there's Katie's habit of landing forward on toes, knees at noon, then slapping down at the bottom of the clock face with heels. Every time Katie landed, the big forces of the ground didn't travel up her body's chain of protective soft tissues. Instead, they zipped efficiently and smoothly through bony heels and up her tibias. At the top of the tibia sits the tender little collection of moving parts called the knee.

No one's biomechanically perfect. We all misplace little forces; over time it can make us creaky. "Kids can jump around and do whatever," says author and elite running coach Steve Magness. "Over time, we narrow our capacity. Without practice, we lose range of motion, fast-twitch muscles, a lot of the things that let us move ballistically." But Katie's P3 assessment didn't just show minor misplaced forces; it revealed that the slap of Katie's heels put her on track for a major knee injury.

P3 would blend Katie's assessment data with insight from medicine, mechanical engineering, biomechanics, data science, neuroscience, machine learning, and athletic training. They'd give her body tools to land differently.

KATIE SPIELER

Katie says she left her assessment with a "game-changer feeling." And as long as she kept up her regular visits, coaches hovering and correcting as she worked through exercises she'd never seen anywhere else, she says she felt "rock-solid bulletproof" on the pro tour.

But everyone knows it's hard to keep up exercise routines. The following year, Katie moved two and a half hours south, to Hermosa Beach, on the far side of Los Angeles, to live with other tour players. Sam Brown left P3. So, instead of going into P3 all the time, Katie did many different workouts.

Katie bubbles with energy, turning heads on the tour for beekeeping, biking everywhere, and ambitious off-day hikes. Her most common kind of Instagram posts are of dancing—on beaches, on volleyball courts, on stage, on football fields, on a rock next to a mountain lake. Then she started dating professional beach volleyball player Eric Zaun, who seemed to have even more energy. People called him "Road Dog" because he lived mostly in a Sprinter van and forever zipped off to explore tide pools, mountaintops, or thrift stores. They're in videos together playing volleyball in the snow, goofing around, and cracking up.

Then: tragedy. In 2019, Eric took his own life. Eric's death comes up often—somehow, Katie extracts a positive message. She talks convincingly about the importance of putting aside distractions to savor special moments.

In June 2021, teamed with her taller cousin Torrey Van Winden, Katie came in third in a giant tournament in Austin—one of her best finishes. But Marcus says some P3 staffers didn't like how she was moving.

Forty-five minutes northwest of Philadelphia, Pottstown is the scrappy kind of place that calls its annual professional volleyball tournament the Rumble. The Rumble has a beer tent at center court and "old-school" volleyball rules—and, instead of sand, this tour stop is played on grass.

"That," says Eric, "is an interesting wrinkle." Katie's risks came from landing; grass is harder than sand.

"I got a tight set to the net," Katie remembers. She was coming from the right side, jumping to her left. "I landed only on my left leg, and I landed on the outside of my foot. And my knee just kind of buckled. And I heard three pops."

Katie yelled. The word she uses now is "excruciating." For thirty seconds, it was unbearable. "Is this my career?" she remembers wondering, at age twenty-six. "I was so, like, not satisfied with my career at that specific time and it was so heart-wrenching." And what about everything else she did for fun?

But "it was the weirdest thing," Katie says. "It was, like, thirty seconds, and then really no pain." By the time staffers crowded around, she could barely feel it. "It's kind of a grassroots volleyball tournament," Katie says, "and so they didn't have a full medical team on site, and they did a few tests and then they're like, 'Oh, so do you want to get back out there?' "

She knew from the popping sounds that she was nowhere near okay. She accepted what help she could get—a small bag of ice—and hobbled around until the end of the tournament. She caught a ride to the friend's house where she was crashing, then flew home the next day without a scan, consult, crutch, or brace. "Getting through the airports," Katie says, "was a little bit intense." Her fourth-floor walkup in Hermosa Beach suddenly made no sense, so she went to her parents' in Santa Barbara to line up medical care.

The accident was on June 24, 2021. It took until July 8 to get an MRI. Like Alex Morgan, Sheryl Swoopes, Megan Rapinoe, Sue Bird, and 200,000 American athletes each year, Katie Spieler had torn her ACL. For reasons that are hotly debated—wider hips, weaker hamstrings, hormones, and neuromuscular issues all come up—women have a two-to-eight-times higher risk of ACL injury than males. Even more maddening: there's incredible research, from different decades, sports, and continents, that simple, often free, prevention programs are incredibly effective, cutting the injury rate by 64 percent. The gist of that work is no surprise to Katie or P3: teach people how to land.

After surgery, Katie found a physical therapist and pushed the limits. Within weeks, she posed on the back porch with her cousin and sometimes volleyball partner Torrey. They wore matching pink skirts, matching white shirts, and matching black left knee braces. Amazingly, Torrey had torn her left ACL at the same time. (The only difference: Katie stood up a step, because Torrey is six foot three.)

Four months after surgery, Katie ran into Marcus. She said ACL recovery was a grind. Marcus felt an almost religious need to restore the healthy free movement of a fellow ocean swimmer, mountain biker, and explorer of the wilds. "He said, 'We gotta get you back in there' at P3," Katie says.

So, she went. Katie says one of her favorite moments of ACL repair was when P3 coaches directed her to stand, in an athletic position, on one leg, with her eyes closed, for thirty seconds. "I'm like, 'Oh, that's got to be so easy.' You know, you're standing on one leg. You don't have anyone pushing you. You're not jumping. And it's so hard."

For most people, vision is a critical part of balance, and balance is a critical part of landing. But volleyball often requires—think about how Katie tore her ACL—a nice, clean, balanced landing while your vision is on the ball. Eyes closed on one leg gooses up your other balancing systems. "The first time I did that, I was wobbling all over. And then you do more and more and you just are quieter," says Katie. "You're stabilizing."

Katie hears often from parents who say their kids get super-fired-up to see someone five-foot-five on the pro tour. She wants to win for them. P3, for her, has not been about improving data. "It's helped me in my sport, and it's helped me in injury prevention," Katie says, "but it's also helped me in my life, because I'm like, 'Oh, I can do this. I can get through hard things.'" Katie spent some of her recovery standing, dancing even, on the smooth orange rocks of the San Marcos Foothills Preserve.

In early May, nine months after surgery, on a sand court in Austin, Texas, an official called out, "All right, ladies, take the court please, take the court please." Katie, in sunglasses and a massive smile, appeared first. "Please Don't Stop the Music" blared. A couple of minutes into the match, Katie served, threw her repaired left knee into the sand and dug out a hard-hit return, stood, planted two feet, and leapt to attack a set from her partner, Kim Hildreth. On defense, in perfect position with arms fully extended, was six-foot, one-inch Carly Skjodt. Katie spiked it right between Carly's arms, then screamed in triumph.

# 7

. . . . . . . . . . . . . .

# JAZZ

. . . . . . . . . . . . . .

*I think I was supposed to play jazz.*

—Herbie Hancock

IN 1997 THE Utah Jazz hired a new director of player development. Mark McKown drove his car to the ritzy Avenues neighborhood of Salt Lake City to visit the last house before the Wasatch Mountains. Karl and Kay Malone lived in a home designed for giants: a taxidermied black bear, a ten-foot-square bed, a chandelier of antlers, a shooting range, and seven-foot-high closet rods.

Karl had a reputation. The Jazz drafted the power forward in 1985, and soon he was paired with John Stockton on a perennial contender best known as the other team in Michael Jordan highlights. Despite his "aw shucks" country demeanor, Malone sent many of the best players in NBA history—Michael Jordan, David Robinson, Isiah Thomas, and Steve Nash—crashing to the court with vicious elbows. Malone's Jazz decades also featured rumors, which he denied for many years, about his having fathered a child with a thirteen-year-old out of wedlock while he was in college.

McKown, a six-foot, seven-inch, shiny-headed bald white guy, had narrower concerns. Mostly: Malone's conditioning, which oriented around long mountain forays. McKown hoped to add basketball movements, especially anaerobic, explosive work, to the mix. It wouldn't be easy: Malone, who had a TV show where he told people how to exercise, wasn't always in the mood to receive advice.

McKown resolved first to earn Malone's trust as his workout partner, and romped over mountains with Malone every summer, all over the globe, for more than a decade. McKown became the kind of backwoods companion who suggested sprinting to the next crest; lateral sliding through the flat section; and then ankle flips—a way to strengthen the lower leg by bouncing off the ball of the foot—through the downhills. McKown never convinced Malone to wear a weight vest, but Malone did eventually lug along a backpack full of rocks.

They also wrestled. McKown called it "functional strength work." Malone called it "a good, old-fashioned ass-whuppin." On a trip to remote Alaska, McKown handed his camera to Karl's brother Terry. McKown's plan was to pose as if he'd finally pinned Karl, who was napping at the time, while Terry snapped a triumphant, if misleading, photo. Instead, McKown has a photo of a very awake Malone smothering McKown. "Terry's timing," McKown says, "was a bit slow."

Malone finished his career as a portrait of NBA durability. In the top ten all time with almost 1,500 games played, Malone retired, body remarkably intact, at age forty.

Then the Jazz needed a new power forward. Rafael "Hafa" Araújo had a couple of strikes against him. The Brazilian Hafa had been banned from international competition for two years after testing positive for the steroid nandralone at the 2002 FIBA basketball world championships—where he also got into a shoving match during a game in which he didn't even play.

More importantly, though, Hafa was a bust. The Toronto Raptors had drafted Hafa with the eighth-overall pick, then traded him to the Jazz a few years later, with cash, for two lightly used players. Leaden feet made it hard for Hafa to do his main NBA job, which was to make it hard for opponents to score. He was strong as a bull, but like a bull, Hafa struggled to move laterally.

Hafa's agent had an outside-the-box idea: send Hafa to the place in Santa Barbara that had helped his baseball clients. And that's how Hafa became P3's first NBA client, coincidentally about the time Marcus started promoting P3's services by handing out T-shirts that said, "No, I don't use steroids . . . but thanks for asking" and "STBD," which meant "Smart Training Beats Drugs."

Marcus didn't know about Hafa's failed test. But years later, the first thing Marcus says is "my eyes tell me" Hafa did steroids. "He got super-strong in college, and couldn't turn his neck."

McKown remembers Hafa introducing him to "his guy" Marcus. At first, McKown was annoyed. Even his team's most middling player had support staff? Then McKown talked to Marcus and found him smart. The third emotional stage was panic; as McKown walked away, he glanced at Marcus's business card and saw he was an MD. McKown played back all the things he'd said, and wondered if he had sounded stupid.

The fourth step in McKown and Marcus's relationship came when Hafa arrived at Jazz training camp a changed man. "It was great," McKown says of the change P3 had inspired. "He could move so much better."

McKown called Marcus. The words he recalls using were "Can I use you as a resource?" McKown says he asked Marcus so many questions about force plates and hip stability that "I think I pissed him off! He asked me if I could just write all my questions down."

Marcus confirms that McKown "had an annoying number of questions."

"I want to give him a lot of credit," says Marcus. "He was scared as hell. He didn't know any of this stuff. But he was not overly affected by his fear."

McKown was learning that a basketball player's movement could be decoded. How was Hafa moving laterally? *You could see it* in the vectors of the force plate.

A key Jazz player came in with a history of groin and hip problems—the Jazz staff had him working on his hip mobility, but it wouldn't budge. McKown says Marcus, "with his eyeball, not an X-ray," suggested a scan to assess the shape of the player's femoral head. Sometimes a hip's limits are soft and movable; other times they're bony and nonnegotiable. When an MRI from a specialist confirmed Marcus's hunch, McKown rearranged the player's training.

McKown loved what he was learning. "If someone tells you they can predict injuries," says McKown, "they are lying out their ass. But can you say there's a probability, or an increased risk? That's what Marcus does."

Soon McKown talked the Jazz into cutting a deal with P3. The following summer, McKown visited with two players finishing their rookie seasons. Ronnie Brewer, Marcus remembers, had a "powerful, electric body." Growing up athletically gifted is, Marcus says, a "double-edged sword," like learning to drive in a new BMW. It's wonderful and smooth, Marcus says. But the clock is ticking until you have a high-maintenance older BMW. Marcus says Brewer "never learned how to fix a tire. When they get a flat, it just stays flat, then riding on rims, bent rim, then an axle."

Paul Millsap, on the other hand, came in with a body like a used Toyota. "He wasn't gifted physically," says Marcus. "He didn't have long arms. He didn't jump very high. He didn't run fast. He looked like an old man. He had achy knees and back, and would be creaking around for a half hour before he could start playing." It was tough to find anyone confident that Millsap would stick in the NBA. Millsap had led the NCAA in rebounding at Louisiana Tech, but didn't create off the dribble, didn't set scoring records, and didn't drop silky passes through traffic. McKown says the Jazz front office didn't even want Millsap—he was on the team because coach Jerry Sloan insisted.

Millsap had a secret weapon, though: his mom, Bettye. With four sons, three jobs, and no co-parent, Bettye developed an infectious brand of toughness. Bettye dismissed Paul's long odds and moved the family to Utah. Every single one of her sons eventually played professional basketball. Paul had Bettye's killer mindset.

But at six-foot-six without shoes, at a position played by seven-footers, did Paul have the body? That's the question McKown brought to Santa Barbara. Marcus made no sunny predictions.

But wow, was Millsap the king of maintenance. "I love those Millsaps," says Marcus. "He was a worker. Just put something in front of him, he'll get it done," says Marcus. The dominating rebounding of Millsap's college days continued in the NBA, while he painstakingly added better ways to move, score, defend in space, and help his team win despite being among the smallest at his position.

The physically gifted Brewer lasted seven years. Millsap played more than a thousand games over fifteen years and retired in 2022, having earned more than $200 million. Despite his creaky body, one of Millsap's great gifts became durability. In his first decade in the NBA, Millsap never missed more than a few games a season. All the time away from surgeons and physical therapists let him work on other things. Millsap became an accomplished three-point shooter and a vastly improved passer.

"I'm convinced," McKown says, "that Paul's career lasted as long as it did because of P3." Bettye Millsap is now the business manager at the Paul Millsap Foundation.

McKown's offseasons became a story of flying to Santa Barbara with Jazz players. Derrick Favors had battled back pain; Rudy Gobert arrived with a clunky, injury-prone body that P3 helped turn into a four-time defensive player of the year; Alec Burks used P3 to recover from injury while dramatically improving his lateral movement. Kosta Koufos flew in after a curious hamstring strain. P3 found that his kyphotic upper back had inspired a corresponding pelvic tilt—he was curved a bit, like the letter *C*, and the tilt made his hamstring vulnerable.

McKown came to see P3 as an extension of the Jazz. "I took it personally," he says, "if I found out they were working with other players." McKown had the Jazz install force plates almost identical to P3's. If they saw out-of-whack measurements, they'd call Santa Barbara.

"Marcus's position," McKown says, "was this shit's great, everybody should be doing this." McKown's position was he wanted to keep P3 a secret from rivals, because "fuck *them*." McKown had human nature on his side. Most NBA trainers were naturally skeptical.

"People don't like to admit somebody does something better, in life in general, if it's in your same profession, right?" says McKown. "And then, if you're in a profession where you see your colleagues get fired all the time . . . you're not even rational sometimes." McKown says every-one in the NBA is "job scared."

Before too long, McKown showed up with Jazz guard Kyle Korver. The grandson of a California pastor named Harold and the son of a pas-tor named Kevin, Kyle is devout in Christianity and hoops. Add a thick coating of Iowa upbringing and Kyle emerges as breathtakingly earnest, a six-foot, seven-inch beacon of straight shooting.

In his senior season, Kyle led Creighton in every major statistical category, especially by making 48 percent of his three-pointers. But his knees hurt. Kyle followed trainers' advice. He iced and did various exer-cises. But a huge part of being an elite athlete is gritting things out, and mostly Kyle resolved to shoot the lights out, make the NBA, and then trust the magical healing hands of an NBA training staff. The first part worked: The first pick in the 2003 draft was LeBron James. Fifty picks later, Kyle.

On draft day, the New Jersey Nets traded Kyle to the 76ers for cash. They told him they used the money to pay for their summer league team and a photocopy machine. Kyle joined a 76ers roster that looked a lot like the NBA Finals team from two years before, but with more scar tissue. Allen Iverson missed half the season; Derrick Coleman and Glenn Robinson missed the poetry and bounce of their younger years. Everyone needed miracles from the training staff. As the youngest and lowest-paid player on the team, Korver read the room and tried to be low-maintenance. And anyway, the trainers' advice for Korver's knees was similar to what hadn't worked at Creighton.

The good news was that Kyle played a little as a rookie and a lot after that. Kyle finished near the top of the league in several shooting catego-ries. The bad news was that his knees hurt in profound ways. By now he had seen, and felt, why NBA careers are so short.

In Kyle's rookie season, one of the players he didn't get to play against was Grant Hill, who sat out all eighty-two games with ankles that were a lot like Korver's knee. So many faces wincing, so many joints in pain. The league was inflamed.

This realization exposed the grave reality of Korver's profession. The NBA employed some of the world's finest athletes. But was that because the league knew how to get the best out of players? Or was it because broken bodies were easily replaced? NBA careers end constantly. You'd be with Kedrick Brown at some shootaround or on the next training table down from Rodney Rogers, and then *poof.* Six of Korver's teammates had careers end in his short time in Philadelphia.

Four and a half years into Korver's career, a talented young Jazz guard named Gordan Giriček said the wrong thing to Coach Sloan in Salt Lake City. Just after Christmas 2007, the Jazz shipped Giriček and some draft picks to Philadelphia for Kyle.

Kyle didn't tell McKown he had secret fears of medical retirement. He told McKown he had sore knees. McKown booked two tickets to Santa Barbara.

Kyle was nearly a decade into the mystery of his knee pain. The P3 staff put him through a full assessment, with its hip range-of-motion measurements, the exploding sideways drills, the dowel hops. Then they showed him the worst movie ever: grainy slow motion of his knees.

Korver recognized his legs, his sneakers, his feet lingering on the edge of the box, and then dropping down to the floor, bending into the landing and . . .

"I almost threw up in my mouth."

Maybe an hour after walking into P3, Kyle met his nemesis. In slow motion, it happened a little, a little, and then more: the knees dipped inward. Finally, near the bottom of the jump, violence: one knee actually crashed into the other.

Kyle had been doing this on national TV for years and nobody had noticed. His right knee hurt because his left knee assaulted it every time he landed. He'd later learn the repeated crashes were but one cause of his pain—the others had to do with the stress of bending inward.

Kyle started listening to every single thing the P3 people told him. He squatted, he pogoed, he became a professional at plyometrics. He learned about *hip abduction*, which is not someone kidnapping your hip, but the muscular action of moving your legs away from the midline. The queen of the hip abductors is the gluteus medius, which attaches to the outside of the top of the femur.

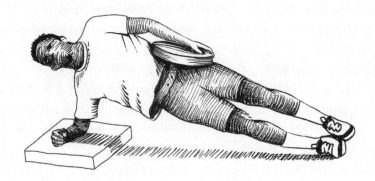

SIDE PLANK WITH WEIGHT

The most direct, brute-force exercise to strengthen the gluteus medius is the side plank. The normal way to do that is to lie on your side, raise up onto a forearm, and then, with core tight, lift your whole midsection off the floor, so your body is straight and your mass is shared between the outside of one foot and the forearm. You might hold it for thirty or sixty seconds. It's demanding even before you add tricks like raising onto a hand instead of the elbow, or lifting the upper leg.

Then there's P3's special variation: weights. Kyle wasn't at P3 long before a coach balanced a weight-lifting plate—fifteen pounds, later thirty or more—on his raised upper hip. In time, he'd grin through the strain as a P3 coach shoved down with two hands on Kyle's hip as he side planked.

A few workouts in, Kyle walked out of the gym feeling like he was made of iron. Hell no, he wasn't retiring. Kyle and his wife, Juliet, bought a house in the Santa Barbara hills with an ocean view. He arranged a gym where he could practice shooting. He spent every offseason and as much time as possible getting stronger at P3.

In 2010, Kyle set the NBA's all-time record for three-point shooting accuracy, making 54 percent. Far from retiring young, Korver finished in the NBA's top fifty all-time in games played. He finished five years as an Atlanta Hawk at age thirty-five, then the Cleveland Cavaliers brought him in to draw defensive attention away from LeBron (the only member of Korver's draft class to play longer), a strategy that twice took them to the NBA Finals. Kyle's NBA tour brought him back to the Jazz in 2018, where he reunited with McKown and somehow made the highest salary of his career at age thirty-seven.

That same year, McKown got his own lesson in being job scared. First, he heard the Jazz had added a new member to the performance team. Then he learned in a roundabout way that the new guy was his boss. Soon he discovered much of the team's player assessment equipment in the dumpster. Within a few weeks, he was informed he was being let go and that the Jazz were done with P3.

"People don't realize," McKown says, "how effed-up these teams are, the politics that takes over everything. It's such a strange business." He decided to retire and moved beyond cell phone service on a lake in the Rockies.

Juliet started wondering if it might be time for Kyle to stop working so hard, to live in California year-round and enjoy some of his $80 million or so in career earnings. When the Suns waived Kyle in 2019, it seemed like time for the next phase. But that summer, Marcus ran into Juliet. Another team had called. She was packing up the family for a winter in Milwaukee. "This," she told Marcus, "is all your fault."

When Kyle finally retired, he moved to Atlanta for a job in the Hawks' front office. Marcus and Kyle see each other often. On Marcus's visits to P3 Atlanta, he stays with the Korvers. When Kyle's in Santa Barbara, he stays in the guest house above the Elliotts' garage. Whenever they're in the same city, they'll find a chunk of time for a wildly intense tennis match or something in an ocean.

They also meet every summer for an insane outdoor challenge. It began when Korver was playing. Marcus was forever identifying some mountain he wanted to climb, some islands he wanted to paddle to, or a cliff he wanted to jump off. P3 clients seldom join—but Kyle was an exception.

In 2013, Marcus told Kyle that it was time for his first misogi.

# 8

## MISOGI

*Much more of the brain is devoted to movement
than to language. Language is only a little thing
sitting on top of this huge ocean of movement.*

—Oliver Sacks

THE HOUSE WHERE Marcus grew up burned down years ago. But the giant fig tree—one of the biggest in North America, more than a century old—still towers over the clear, cold water of the spring. That's where Marcus learned how to fly.

Fig trees drop ropes like Rapunzel and entice children up. From atop the trunk's shack-shaped base, choose-your-own-adventure branches jut at every angle and elevation. Marcus leapt into the ad hoc swimming pool below. Every year he'd jump from higher, "to the point," Marcus says, "where you knew exactly what you could do."

Baby birds know the story. Pink, downy, eyes-closed hatchlings can't fly. Over weeks, though, the babies get fidgety and flappy. The parents bring food less often. The adult birds, at least in some species, lurk a yard away and appear to do the opposite of what we generally consider good parenting: they call the kids to the ledge. Eventually, the babies teeter on the brink, sleek with new feathers, and give it a whirl.

Not all have the knack—some tumble to the ground, where predators lurk. But every fledgling carries a stew of potential—DNA, mindset, audacity—sufficient to leap off the edge and try to flick on the flight switch. Usually, it coalesces in a miracle.

Perhaps the danger of crashing fires something up in the young ones. "If you missed," Marcus says of the water below, "shit was getting broken. A lot of it." Marcus loved how the danger of free fall honed his attention: "The stakes were so high, your senses got so sharp in that moment." He tuned in to every detail: Leap off one foot or two? Eyes open or closed? Feet parallel or offset? Breath held or exhaled? Butt clenched or relaxed? Legs splayed easy chair–style, crossed, or together?

Officially, humans can't fly. But some are more ready than others. One day, Marcus got confused by something he read in the kitchen, with the fig tree just outside. People *died* jumping off the Golden Gate Bridge? He calculated that, from two hundred feet up, jumpers would hit the salt water at seventy miles an hour. Admittedly, that comes with forces akin to being hit by a car. But free divers compete, intentionally, from almost as high. Marcus had studied the operatic water crashes of ospreys. That predator dives from as high as the Golden Gate, breaks the surface at eighty miles an hour, and doesn't damage a single delicate bone. "You have to shoot out," Marcus says, plateauing his palms down and away like wings. "You hit the water with energy. You have to spread that energy out."

One of earth's great pioneers, 150 million years ago, was an unnamed dinosaur who became the first of her species to fly. Johns Hopkins researcher Amy Balanoff published a groundbreaking paper on this juncture of evolution, and concludes the magic happened largely in the cerebellum, what Balanoff calls the "flight-ready brain." Scans of pigeons show exciting activity in the cerebellum as they take to the air; the fossil record shows that skulls shifted to accommodate bigger dinosaur cerebellums at the time they evolved to fly. Animals that had never flown somehow grew ready in the head.

One term for that is *motor programming*, which Marcus brought up when I texted him a link to a *New York Times* article saying a typical child these days plays only three years of sports and quits by age eleven. Meanwhile, about 2.4 billion kids play video games.

"Yikes," he quickly replied. "Intuition is more compelling than research here. What sucks and is most always missed is kids who don't expose their bodies to challenging physical activity as youngsters don't develop the tools to do so in an engaging way as adults—the motor programs etc. that make moving so damn much fun!"

The words *motor programs etc.* open a porthole. Like language and music, human movement is a product of the mind. A good jump is as brainy as learning Spanish or playing saxophone. Hurdling a fallen log, catching a Frisbee on the run, standing up on a surfboard—these are also things to rehearse, which is often called *play*. As a parent, you want your kids to learn not just how to stand and walk, but really *move*.

And that happens in the brain. We share huge swaths of DNA with animals that do magical-seeming things, and we have astounding capacity to learn.

Daniel Kish lost his eyes to cancer at the age of thirteen months, and then—no kidding—learned to echolocate like a bat. He makes a clicking sound with his mouth; the sounds bounce back to his ears in a way that lets him map the world so perfectly that he can mountain bike and play basketball.

A 2011 study in *PLOS One* watched activity in the brains of Kish and another human echolocator named Brian Bushway. Researchers found that if you play Kish and Bushway any old sound, their brains react typically. But play their own clicks bouncing off things, and returning to their ears, and parts of their brain normally associated with vision— especially the calcarine cortex—light up with activity. They can identify, from recorded sounds of their own clicks, if they're bouncing off a friend or a tent pole. Their brains appear to have new wiring, formed at some point after birth. "Our data clearly show that [Kish and Bushway] use echolocation in a way that seems uncannily similar to vision. In this way, our study shows that echolocation can provide blind people with a high degree of independence and self-reliance in their daily life. This has broad practical implications in that echolocation is a trainable skill."

*Neuroplasticity* is the word. The brain's structure can shift. We constantly create and reorganize synaptic connections, especially in response to experience or learning. Researchers Matt Puderbaugh and

Prabhu D. Emmady write in the academic publication *StatPearls* that "exercise, the environment, repetition of tasks, motivation, neuromodulators (such as dopamine), and medications/drugs" all induce an interesting pattern: the electrical activity associated with moving shifts over time, from the bilateral premotor cortex, which has ready access to the spinal column and a role in planning movement, to another part of the brain, the right hemisphere supplemental motor cortex, which is associated with complex, full-body movements. One theory is that during this shift, movements are being encoded.

Some of the key research into neuroplasticity has been conducted on owls, whose survival hinges on vision. (In some owl heads, eyes comprise a full third of the mass.) Researchers affixed prisms over the eyes of juvenile owls, which meant they would only see clearly by learning to unscramble visual stimuli in real time.

And they did. The owls rapidly sprouted new axons, write José L. Peña and William M. DeBello in the *Institute for Laboratory Animal Research Journal.* The long, threadlike nerve cell parts "represent the construction of a learned circuit," write the researchers. "These experience-induced synapses are very likely the principal source of learned responses." The owls built a workaround. And, the researchers suggest, the findings "may well provide a path toward revealing universal principles of information processing in the brain."

Sometimes animals make incredible leaps forward. Ground squirrels did not hibernate until, one harsh winter, some audacious little furball developed the ability to go months without food. Kish learned to navigate the world by sound after the crisis of losing his sight—but research shows merely wearing a blindfold for a couple of days attunes a human brain to sound and touch.

In her book *Mindset,* Carol Dweck, Ph.D., writes that "no one laughs at babies and says how dumb they are because they can't talk. They just haven't learned yet." In her practice, she sees incredible results from telling teenagers that people are naturally neither smart nor dumb, but instead that "the brain is more like a muscle—it changes and gets stronger when you use it."

Almost two decades after Dweck's groundbreaking book, it's

almost common knowledge that we can influence the makeup of our own brains. What might be fuzzier, though, is the degree to which that insight applies to sports, and how we move.

Think about the difference between a typical right and left hand. They look similar, even by X-ray—nearly identical muscles, size, and range of motion. But basketball coaches take it as fact that a backup point guard could use a summer of practice to bring mastery to her off hand. We know that practice conquers clunkiness.

So much so, in fact, that many NBA players upend the very concept of being right- or left-handed. The term is *cross-dominant*. LeBron James, Russell Westbrook, and Rudy Gobert all sign autographs left-handed, but if you leave any of them open with the game on the line, they'll shoot with the right every time. Jaren Jackson Jr. shoots three-pointers right-handed, but plays lefty when closely guarded in the paint. Coaches have long argued about which hand Ben Simmons should shoot with.

Which makes me wonder: Am I right-handed, or is my left hand merely undertrained? The only people I know who worked on both hands day in and day out through their developmental years are elite athletes, particularly basketball players. Were they good basketball players because they were naturally ambidextrous, or did they become ambidextrous at basketball practice?

Brains scans offer a little insight: compared to the 90 percent of people who identify as right-handed, left-handers tend to have a larger corpus callosum, and increased activity in Broca's area and the planum temporale. But lefties aren't the only ones who can develop these parts of the brain.

The factors in play, according to *Nature*, are "evolutionary, hereditary, developmental, experiential and pathological." In other words, vulnerable to work. Interestingly, these brain areas are also all involved in processing language; *fluency* is the word applied both to linguistic and physical mastery.

That's what Marcus wants people to work on. Mastering movement is a little like conjugating the verb *hacer* or soloing like Ornette Coleman. They're all about neuromuscular coordination. Which is great news: we can gain command of wayward hips, knees, and ankles. Marcus routinely finds himself talking to athletes like Kyle Korver who have some oddity like knees knocking together on landing, say, which might seem beyond their control.

He's often asked some form of the question: Can this really change? The answer, Marcus says, is yes. Always yes. "*All* those things can be changed." Because the brain tells the story of how we move, and brains can change.

David Grann makes a case in *The New Yorker* that squid may be the best movers on earth. He quotes a scientist as saying that the only squid we can catch are slow, sick, or stupid. No one has ever caught a living adult giant squid, which is the point of Grann's story. Just as flying lights up a bird's cerebellum, the squid's genius seems to stem from its nervous system. Grann writes that squid have nerve fibers that are "hundreds of times thicker" than human nerves. Much of the research into those tissues has focused on the role of myelin, which is something like insulation for brain circuits. As described in Daniel Coyle's book *The Talent Code*, practice can increase myelin, which essentially upgrades the circuitry. Things flow faster.

Squid must constantly detect and elude master predators like sharks. Evolution has tested the squid's neurons, axons, myelin, latent DNA, and everything else. It's as if they have clamped the brain's jumper cables onto ancient and torrential forces of movement improvement.

Marcus and his medical school buddy Garth had ideas about sprucing up their own neurology by staring down big, physical challenges. In 1993, while in medical school, Marcus flew from Boston to Wyoming. The plateau in the middle of the Wind River Range is at 12,200 feet—about double the height of the tallest mountain on the East Coast, and higher than revered Western peaks like Mammoth or Mount Hood. The plan was to crisscross the range for more than a week, carrying the minimum, catching most of their food, and fighting altitude sickness with melted snow and gritty resolve.

An elite practitioner of a variety of Japanese martial arts, Garth enthralled Marcus with discussion of the Shinto religion, which holds that there are supernatural spirits in everyday things like birds, trees, and plants. "It is hardly necessary to say that it includes human beings," wrote Motoori Norinaga in his 1822 text *The Spirit of the Gods.* Then Garth described a Shinto tradition of cleansing and purification called *misogiharae*, which includes intense ritualistic challenges. Garth and Marcus got excited to see their own trip this way, and started playing with the word *misogi*.

Plans change quickly in the mountains. When Marcus tells the story, it becomes a survival novel. Garth got altitude sickness and flew home. The Rocky Mountain fish refused to touch an outsider's fly. The food ran out. The weather turned hellish. Marcus nearly fell to his death, scaling a cliff in a rainstorm, when a sapling ripped out in his hand. Losing weight and tiring easily, Marcus stood on the bank of a stream where he could almost touch the trout. When they ignored his fly, Marcus considered netting, trapping, and spearing. There are ways to catch a trout barehand.

Then he saw a giant red ant crawling on a log. When Marcus tossed the ant-baited hook into the water, its wriggling legs triggered ancient trout impulses. Marcus feasted until the grease ran down his chin. "There are a few million years of evolution in that," says Marcus.

On the eighth day, as Marcus tells the story, he was singing a song, winding his way back toward the road where he would hitchhike to the airport, when he heard the chuffing of a horse.

"But," Marcus remembers, "I couldn't see a horse."

The chuffer appeared on the trail ahead. A "giant fucking bear," Marcus says, "sniffing so loud, his mouth open, snorting." Marcus acts this out by stretching his jaw as wide as possible and sweeping teeth around in big arcs—like the bear's mouth was a satellite dish dialing in the Marcus channel.

If a grizzly charged, the brochure said, don't act like a fleeing prong-horn. Stand your ground and the bear *might* veer away at the last second. Bears largely eat ants, roots, and rotting elk carcasses.

Plymouth University professor Mark Briffa has spent his entire career researching the "contest behavior" of animals. I asked him what kind of research undergirds these life-or-death government recommendations. "I'm not sure that there is any science behind this," Briffa replied, guessing instead it's built on "a history of trial and error."

Look at the accounts of people hurt by grizzlies, though, and there's a lot of error.

Marcus says he grabbed his knife and looked up the trail. A still bear seems fluffy, cozy, and round. But when it moves, the lean killer emerges—like a boxer throwing off his baggy robe. The grizzly barreled at Marcus, murder quick. Marcus clenched his knife and prayed the bear would veer. In a way, it did, running directly to the top of the huge rock looming above Marcus's patch of trail. "The feeling," Marcus says, "is just a total lack of control."

Just a minute before, he had been singing and life seemed better. After some time, Marcus says, he tried a few notes. And then a few more, a little louder. In song, he ventured one step up the trail, and then another, faux casual as a shoplifter. "My heart rate," Marcus says, "was up for a long time."

But later that day, safely on a flight back to Boston, Marcus felt empowered and improved by having survived the trip's profound uncertainties, like a squid that had dodged a shark. His own brain, he felt, had been upgraded by gripping that slippery cliff, catching dinner, and lullabying a grizzly. At the time, Marcus was training for triathlons by teaching himself to ignore his brain's proposed limits. And he was marinating in cutting-edge physiology and performance findings from researchers like Steven Horvath, Bert Zarins, and Timothy Noakes— all of whom discussed fatigue as a fungible human emotion. What other perceived limits were bogus?

Stanford psychologist Alia Crum has studied the effects of stress on the human body. Writing with her husband, Thomas Crum, in the *Harvard Business Review*, she says that "the body's stress response was not designed to kill us. In fact, the evolutionary goal of the stress response was to help boost the body and mind into enhanced functioning, to help us grow and meet the demands we face. . . . Although the stress response can sometimes be detrimental, in many cases, stress hormones actually induce growth and release chemicals into the body that rebuild cells, synthesize proteins and enhance immunity, leaving the body even stronger and healthier than it was before."

"Real toughness," writes Steve Magness in his book *Do Hard Things*, "is experiencing discomfort or distress, leaning in, paying attention, and creating space to take thoughtful action. It's maintaining a clear head to be able to make the appropriate decision."

After his trip to the Wind River Range, Marcus built a habit of stretching his own brain with absurdly ambitious outdoor missions. On a trip to visit Nadine's family in the Alps, he was halfway up a sheer face so high and dangerous that a helicopter arrived to rescue him. He waved them off and called Nadine to make sure she knew he was okay. Marcus's friend Jonas Jungblut has dozens of stories about missions with Marcus. "It's usually," Jonas explains, "relatively in control."

If Marcus's life's work is teaching the joy of movement, the work that happens within the walls of P3 is akin to language school. *Yo soy, tu eres* . . . they drill the vocabulary and the grammar of hips and ankles, then they send an invoice. That's a great way to learn. But there's also the full-immersion method: your family moves to Guayaquil and you've never spoken Spanish before. *¡Suerte!*

That's *misogi*. That's what Marcus wanted to give Kyle, as a gift, when he suggested they paddleboard from the Channel Islands to Santa Barbara. It's approaching thirty miles. Kyle grew up in the landlocked Midwest, had never paddleboarded one measly yard, and said sure.

Kyle fell off his red-and-white board within the first minute. They had nine hours to go. The wind kicked up. The waves slapped and menaced. In an area known for great white sharks, someone—whoops!—dumped a chicken burrito in the water. And then there was the approaching upright fin of what turned out to be a sunfish. Muscle spasms, bloody toes—Kyle found the whole thing hellish.

Even for a professional athlete, a big workout session might last an hour or two. (Thirty seconds of a side plank can seem like a lot.) The paddleboard trip lasted long enough for some people to run three marathons. And Kyle had not trained for it.

Kyle says he discovered a new way of thinking that day. He stopped worrying about how things might have gone, and dialed his focus tight on the idea of creating one perfect stroke. Get the feet just so, the posture, the grip, the depth of the paddle blade. He made a huge mission small. One stroke.

The following season, Kyle set an NBA record for one perfect stroke after another; he made at least one three-pointer in ninety straight games. Kyle attributes that record directly to the misogi. People who go through a misogi say they feel changed. A lot of players show up to play, Kyle says, to see what would happen. Kyle resolved to be far more intentional. He'd prepare a certain way, and in the epic grind of the NBA season, would dial in his focus to one stroke at a time—feet, posture, and motion just so.

The next summer, for Kyle's second misogi, the shooting guard found a rock that, on his bathroom scale, weighed a little more than eighty pounds. Someone else found a sixtyish-pounder. They tossed the rocks to the floor of the prettiest little cove of Santa Cruz Island, where Katie Spieler ran around in her youth. The method was to take a breath, dive down, wrestle the rock off the murky sand, and then run. Or, at least, struggle a few steps before dropping the rock and ascending for a much-needed breath. Then your partner takes a turn. In the early going, most people can carry it a few feet. They had decided the night before, in the Korvers' living room, that the goal would be five kilometers.

Writer Charles Bethea described the experience for *Outside*:

> The first hour is an eye-stinging, lung-burning, pride-killing exercise in futility. The second hour, too. And much of the third. I get tangled up with the story's underwater photographer at one point and nearly come to blows. Elliott accidentally rakes me across the face with the rock. Shortly thereafter I hit his shin. There are few words exchanged beyond: "Here . . . You got this . . . Good job." Except I'm not doing a good job. I can't seem to take a big enough breath. I'm often gripped with panic when I touch bottom and try to move. I can't get traction. My gloves feel too large, so I rip them off. My goggles fog and I curse them. I bemoan my employment, my employer, my god.

But then . . . the gurgle of self-talk adjusted. Everything remained steadfastly nightmarish, but Bethea began to carry the rock farther. After four hours and forty-nine minutes, the teams made the full five kilometers. Bethea called his mom to say that he felt a little superhuman, to which she replied, "That doctor guy might be full of shit."

Everyone reports feeling tired after misogis, and otherworldly. "We hear a lot about how stress can decrease your cognitive performance," Crum writes on a Stanford behavioral science website. She continues:

> Scientists found that the subjects in the midst of a bungee jump can process information much faster than a non-freefalling control group. . . . When a group of patients was purposely stressed before going into knee surgery, they recovered at twice the rate of a control group not primed with stress. This makes sense from a historical perspective. If you get attacked by a saber-toothed tiger, survival depends on a high-functioning immune system. If you get hit by the tiger, you want your immune system to respond very quickly. This is how vaccines work as well. They stress your body with an overload of antigens to create an active immune response. It's also how we get stronger. Weight lifting stresses our muscles to the point where we break some muscle fibers. As they heal, they rebuild stronger than they were before. In some of our most stressful life events such as battling cancer, being in an accident or going to war can cause huge leaps and personal growth.

Another year Marcus, Kyle, and friends climbed the height of Mount Everest in the stairwell of downtown Los Angeles's Wells Fargo Center. Jonas snapped an image of a grinning, shirtless Marcus flipping the double bird to the camera as a suited man, work ID on a lanyard, reaches to press his elevator button.

When humans go to the moon or cure cancer, Marcus says, success has never been likely. "Every cell says stop, and you keep going," says Marcus. "There's something that happens. Some self-respect comes from that type of quest."

Marcus's notion of the misogi is that you dream up your own definition of a task that you suspect you have about a 50 percent chance of completing. Make it novel. If you're a runner, don't run. There's value in going in a little blind and feeling out of place. No entry fees, no tickets, no cheering audiences. Just you and perhaps a few friends, in nature, stretching at the taffy of the possible.

When Bethea flew home after his first misogi, he didn't know what, if anything, would matter about it. But almost a decade later he sees himself as "a different type of central figure in my story, one who was a little bit more confident, a little more bold, a little bit more capable." Bethea has attempted a series of extreme challenges since. After everyone else in his group turned back, he took a mountain bike to the frozen peak of the 22,000-foot Ojos del Salado volcano in the Andes. He rode a touring bike from Portland to Breckenridge. He captained a twenty-four-hour relay team that won a trail run. He ran a sub-five-minute mile. And, Bethea points out, he became a staff writer at the *The New Yorke*r, "which is also pretty statistically improbable."

"It points back somewhat to the misogi experience," Bethea says. "I don't want to overemphasize it too much. But I think it has to be related, because these things started to pile up in the years following that experience."

Perhaps the most dire physical feat I have witnessed was the birth of my two children. Each time, my wife, Jessica, made it through the early hours in silence, with a peaceful look on her face and closed, meditative eyes. The active pushing and births she navigated with a certain ferocity and, by way of painkilling, not so much as an Advil.

"We all know that having a baby is such a physically overwhelming thing," Jessica says, looking back more than a decade later. I am wondering if she carries, now, wisdom from the trials of childbirth. "I do think that changed me forever," she says. "I feel like it's a superpower." I've witnessed this many times. Shoulder injury, wicked burn, she handles these things with badass calm.

In 2018 Marcus was in Croatia when he got a call that his dad was unconscious, hooked up to all kinds of machines, and it was time for Marcus to make hard choices. "He didn't want any of that stuff," Marcus says. Marcus told the hospital to unplug him, it was over. No one is immortal. And then he says he started crying, running through a forest of giant cedars, "with all these tangled roots." His mother died the same year.

At some point in the grieving, Marcus told Nadine that he wanted to climb the tallest mountain in the lower forty-eight. Mount Whitney is only a few hours' drive away, and a formidable opponent. There have been newspaper and magazine stories about the alarming number of helicopter rescues, deaths, and even abandoned cadavers. (That year, *Outside* called it a "cluster of catastrophe.")

There's a rigamarole to get a summer permit, so Marcus and Jonas set out in late autumn. They left the tent in the dark, early morning with day packs and a pledge to turn around if things got dicey. For Jonas, that moment came while trudging through unexpected snow on a notoriously steep swath of switchbacks. Shortly after that, Marcus happened into some drama around a couple that had spent an unplanned night on the mountain and were lost, cold, injured, hungry, and so close to death that they had already made a video saying goodbye to their children.

As another hiker raced downhill to summon authorities, Marcus expended all of his remaining water, energy, food, and daylight cajoling and carrying the couple to the spot near the summit where a rescue helicopter could land.

When the chopper finally arrived, Marcus used precious phone battery power to shoot a quick video of the triumphant rescue.

It happened so fast, though. One moment, he was celebrating; the next instant, the chopper whizzed off into the sky.

Without Marcus.

"They left him on the top of a mountain, at night," says Nadine. "I'm from the mountains; I know that's not okay."

Marcus faced a vicious form of misogi. On a clear day, the hike down would be a tricky half day. In the dark, the only safe move was to seek shelter. But Marcus didn't even have a sleeping bag. So, he took off his hiking boots and put on the trail running shoes from his backpack, filled his empty water bottles with snow, and took off running. When the daylight petered to nothing, he ran by headlight battery. When that ran out, he spent precious phone battery on the flashlight.

The further he descended from the summit, the more Mount Whitney spidered out into rocky ridges. Each canyon looked beautiful, but only one led to the tent with food, shelter, and Jonas. Low on everything—sleep, food, heat, light, water, oxygen, battery—Marcus found himself at a fork he simply could not recall.

In the tent, Jonas calculated how long it might take Marcus to climb the mountain, and then adjusted everything for the possibility he'd been roped into helping the stranded couple everyone was talking about. He spent a lot of time gazing uphill, hoping to see Marcus steaming down the switchbacks. When Jonas finally decided to call 911, he had no cell reception. So he decided to hope Marcus would arrive; then, if necessary, he would seek out the ranger in the morning. "I didn't like it," says Jonas.

On the dark mountain, Marcus thought through his situation in a way that would later really impress Nadine: he checked his photographs. Almost every snap on Marcus's phone is of outdoor adventure: his daughter Mila balance-beaming a log over a ravine; his other daughter, Kira, and Nadine on a run; hugging his son, Keean, by a snowy lake, and . . . sunrise with Jonas, halfway up Whitney that morning. In the background, Marcus could make out just enough of the horizon, the tree line, and an outcropping to see that they had ascended to the left of this ridge, and the right of that one. Triangulating, he picked his way into the correct canyon.

Jonas remembers it was almost 11 p.m. when Marcus came "crawling into the tent, just destroyed." Jonas tried to give him food, but Marcus almost immediately fell asleep. The next day, two things started: a big storm, and Jonas and Marcus's ongoing disagreement about how close Marcus was to death.

Marcus says he had a backup plan: if he got lost, he was simply going to climb the 14,502-foot mountain again, in the dark, and spend the night in the uninsulated stone hut astronomers built in 1909, where he'd do jumping jacks all night to stay warm.

"I don't think he would have made it out," says Jonas. "He usually looks pretty put together and you can see the confidence in his face. He had the eyes that told you the story of not being okay. He was sort of shaky, and, yeah, that was a rough one."

# 9

## MOTION CAPTURE

*Nothing is more revealing than movement.*

—Martha Graham

"WHEN YOU WATCH an animal move, when you watch an athlete move, and they move well, what you see is poetry," says Marcus. The results P3 had achieved with force plates encouraged Marcus's suspicion that the dynamics of elite movement were immensely complex and underserved by the academic journals. The gold standard of academic biomechanical research is what's known as a *univariate analysis*. By and large, to be published, biomechanical findings must be reduced to one factor—meaning that journal articles squash the complex lyricism of a jumping human. "We take that poetry," Marcus says, "and we cram it into a couple of columns. Now we have a two-by-two table. 'He flexed this much, he didn't flex that much. Is there a relationship between jump height, and how much you flexion your hips?' That's how we look at these things. And that's forcing it into a system that is so far from how that poetry should be read."

The essential output of the force plates is a graph called a *force-time curve*. It shows how hard an athlete pushes, and in what direction, with each foot, over the period they're on the ground. Most athletes fit a pattern of pushing with a certain number of newtons at five milliseconds, and then double that at ten milliseconds, and so on in a manner that Marcus calls "super-linear." But a special group—in Marcus's estimation, the best movers—land and jump with force-time curves that look like ski jumps. Marcus says they "create these exponential curves that are more indicative of high power output, where the rate of force development is accelerating as they go deeper and deeper in a movement. The curve is steeper and steeper and steeper until it's vertical at the end."

When Marcus saw the data, his first thought was: *Okay, we got something here.* It was a hint at hidden layers of athletic brilliance. "But only up to a point," says Marcus. "You're really kind of guessing what the whole system is doing." Marcus had questions, theories, and a commitment to the scientific method. "You don't know," he says, "until you measure." They would need a new tool.

"Usually," says Marcus, "the way these things happen is you have a hammer, and you go look for a nail, right?" You know what makes doctors recommend ultrasounds? Purchasing an ultrasound machine! The same goes for skin lasers, AlterG treadmills, health-tracking software, and a million other things. Labs tend to be built around some technology. Marcus wanted to build a lab around human movement. "We were just looking," he says, "at the nails."

For a few years, UPS and FedEx paraded into P3 with a steady stream of the industry's new favorite gizmos: wearables like Catapult, Whoop, and later the Oura Ring and Apple Watch (Marcus calls them "inertial sensors"). The little wearable doodads could never answer his questions. "You mostly get that the athlete was going this fast for this long, which would be interesting if you wanted to know about energy expenditure," Marcus says. "I'm kind of amazed they took off as much as they did, because nobody's doing anything with them." A 2020 survey of available research sounded notes of optimism but struggled to identify real scientific breakthroughs from wearable devices. Even on the simple topic of which player is tired, Marcus says "I don't think they're as insightful as a coach that's really paying attention."

P3 needed more robust answers, which led them to consider something that "wasn't even on the menu" when he started P3: motion capture. "It's super-slow," says Marcus. "It's super-expensive. It's not very sexy. And the data you collect was very granular, really very complicated."

The work was so arduous that in Marcus's view, the field had been repelling interesting thinkers for decades. On top of that, there was no opportunity to own the technology involved, as the basics of it—a ring of cameras aimed at an athlete—had been the same for three-quarters of a century.

On the other hand, the force plates stirred up questions motion capture might answer. "That's always how it works: every time you find an interesting answer, then you have more questions, right? If you have a curious mind," Marcus says. "So that's how we got into motion capture." Precisely when the whole world was diving into quick, free, and streamlined data collection from wearables, P3 would pivot to a system famous for being slow, expensive, and complex.

Looking for someone to lead the effort, Marcus found Eric Leidersdorf. "This is probably gonna sound weird," Marcus remembers Eric saying in the job interview, "but I've always wanted to be a world expert in something. If I had to choose anything, it would probably involve basketball."

Eric finished a biomechanics degree at Stanford in 2011. Out of college, he interned at the shoe company K-Swiss and worked at the sports apparel company Oakley—where he had helped design an incredibly lightweight golf bag. In 2012, after Eric got a little bored, a coworker suggested he reach out to this little biomechanics shop in Santa Barbara.

Six months after he sent Marcus a letter, Eric's phone rang with a Santa Barbara area code. Soon after, Eric drove the hour to P3 from his parents' house in Thousand Oaks in his hatchback. "There was a Bentley and a Rolls parked out front, I'm parking my beat-up 2005 Subaru Outback. As I'm walking in, Al Jefferson and Andris Biedriņš are walking out." The shorter of the two NBA players is six foot ten. Eric is five-nine.

To Marcus, many job candidates lacked scientific rigor. Then there was Eric, who lost sleep over the replicability of research, never claimed anything he didn't have the data to support, and had an ease with phrases like "helping athletes shift those probabilities," which was exactly what Marcus wanted to do.

When Eric joined in 2012, P3 changed. It had been a bespoke gym, expressing Marcus's medical approach to fitness. Now it would be a sports science lab.

After a thorough review of motion-capture systems, Eric settled on a company called SIMI. He jokes that he still has spider bites from the time he spent in the rafters, installing and calibrating the infrared cameras with a group of friendly Germans. The system is beyond persnickety. There is a reason, Eric says, that motion capture–based academic papers tend to have tiny sample sizes: Who has the time to scan hundreds of people? The athletes must move freely, which means unpredictably. But at least two cameras must see every part of an athlete's body at all times so they can triangulate locations. So, you're aiming cameras very precisely at no point in particular. P3 staffers calibrate and recalibrate to this day.

On April 5, 2013, Eric wore khakis to work and watched as SIMI's North American sales rep, a guy named Benedict, prepared one of China's best young players, seven-foot Wang Zhelin, to enter P3 history. Wang had been touted as the next Yao Ming after scoring nineteen points in the 2012 Nike Hoop Summit. Wang is largely forgotten now, except at P3, where he's famous as the first person ever assessed with their 3D motion capture.

The scan didn't go well. The software mocks up a live animated skeleton that bounces around the screen as the player jumps and cuts. A few seconds into Wang's assessment, most of his skeleton blinked, then disappeared. For a time, poor Wang was "just a floating pelvis." Watching a replay, Eric says, "That doesn't look healthy," and I'm not sure if he's talking about the motion or the capture.

After any assessment, the team must arduously hand-correct every frame, to assure the cameras are seeing what they think they're seeing. In the early days, it could take as long as twenty hours to process one assessment. Marcus says that P3 would pay people to work fourteen hours a day—sometimes as many as six people at P3, and more at a vendor in Serbia. P3 was "just burning money," Marcus says, "like, for years."

And it's not as if motion capture attracted business. "I'm very confident," Marcus says, "that there were zero people asking me for this."

But with each assessment, torrents of human movement data rolled onto P3's servers. "It took a long time," Marcus says, "before we had enough data to have confidence that we knew anything."

In time, insight blossomed. Two decades into his quest to prevent injuries, Marcus had found the tool that would unlock everything: a twentysomething named Eric running the SIMI.

"These are great moments," Marcus says. "Some of my favorite moments." For example, they discovered that players had movement signatures, almost like fingerprints. If you asked ten NBA shooting guards to complete a drop jump, they'd each do it differently. But they'd each do it the same way, or close to it, every time—even months later.

The idea that athletes have signature movement patterns could matter in addressing one of the key issues in sports injuries: determining when a player is ready to return from injury. The SIMI data offered a new question: Is the player moving like *himself*?

At the time, the models were pretty basic: you were considered recovered from ACL surgery if you could bend your knee 120 degrees four weeks after surgery, or press a certain amount of weight with your quad. Those simple rules applied to NBA point guards, train conductors, and retirees.

But one of the most reliable predictors of future injury was past injury, which suggested that the return-to-play protocols had flaws. Marcus sometimes observed bodies compensating for injuries. People often told Marcus that when they hurt their right knee, they had been worried about, say, their left ankle. "A lot of times, your intuition is right," says Marcus. "They can tell they're doing something that could be dangerous on that day.... What if we had better information on those systems?"

One way to assess a player's readiness would be to compare a post-injury assessment to a pre-injury assessment. "What if we had Hansel and Gretel breadcrumbs," Marcus asks, "that you can sprinkle through the forest that allow you to get back out? Instead of just guessing?"

These questions, habits, and movement patterns had long danced fuzzily on the horizon of Marcus's thoughts. Like Eadweard Muybridge, who used advanced photographic techniques to answer pressing questions of movement in the nineteenth century, Marcus and Eric wanted to dig into something fundamental. Before Muybridge, people thought horses ran like leaping cats: front legs stretched forward, back legs trailing behind. Others thought horses kept at least one hoof on the ground. No one knew for sure. Human eyes were simply too slow to grasp how horses actually ran—but cameras weren't. In July 1877, Muybridge solved the riddle by making the first convincing still photographs of galloping horses. He showed that horses did, in fact, leave the ground—and popularized the concept that seeing things every day wasn't the same as seeing them well or clearly.

Eric's idea was to use motion capture to really examine how we jump. "We had our battery of tests kind of lined up for the launch run, one of which was on the drop jump," Eric says. The drop jump traditionally asks the test subject to step off a foot-high box, land on the ground (or, in testing, often a force plate) on two feet, and then explode upward.

DROP JUMP

At P3, it's slightly different. The athletes start on an eighteen-inch red-and-black Retrospec-brand plyo box. It's six inches higher than the standard test because P3 figures highly trained athletes can handle the impact, and with slightly bigger forces, P3 can expose more of the athlete's movement signature.

Eric's target, the drop jump, happened to be one of the most studied movements in sports. Not only was it all over the scientific literature, but it's also one of the most universal movements across the animal world.

From a kangaroo rat to a house cat, almost every ambulatory animal relies on jumping in some way. Mountain lions can jump fifteen feet, vertically. The highest recorded animal jump was a dolphin, leaping almost twenty-three feet in the air—high enough to clear the biggest giraffe in history, if you could get a giraffe to stand on an ocean raft.

Humans use many different systems to jump. The main theme is to bend the major joints, drop the hips, and then contract many muscles at once to rise in the air. Three of the four quadriceps known as the *vastus muscles* generally do the most work, followed by the hamstrings, gastrocnemius calf muscles, and glutes. If those combined efforts can defeat Newton's laws of gravity, you'll get off the ground. From the ancient Greeks, who had jumping in the original Olympics, to the NBA dunk contest, humans everywhere have tried to figure out how to jump higher.

Ads in the back of sports magazines in the 1980s said elite jumping came from unusual and expensive shoes that kept your heels from touching the ground. A couple of decades later, for a magazine story, I worked intensely with an NBA team trainer—Greg Brittenham, then of the Milwaukee Bucks—to learn to sky like dunk champion Desmond Mason. (Brittenham outlined a detailed program built largely around Olympic-style lifts.) Former NBA player and coach Don Nelson recommended playing basketball on snow-covered courts in Wisconsin. David Epstein writes about the magical powers of a good Achilles tendon, which can propel humans, kangaroos, and other animals into the sky.

Thanks to their new toys, the team in Santa Barbara could see this old conundrum literally through a new lens. They didn't have the *first* motion-capture system in the world, but they had the first one capturing data, in bulk, from extraordinary athletes. Eric says "the responsibility was there, at that point, for us to start digging into it."

What were the characteristics of a killer jump? What mechanics led to good height? At first, Eric thought these questions would be "easy to study." They didn't think crazy high jumps especially mattered—in every study, jump height simply doesn't correlate with NBA success. But P3 did have a lot of insight.

"We have this output, jump height, which they measured at the combine for forever in different tests," says Eric. "One output. But we've got these hundreds of variables behind it. Let's see what, of those, ultimately contribute to that jump height."

P3 had several years of force-plate data from NBA players like Shabazz Muhammad, Rodney Williams (who rated as one of P3's finest-ever athletes), Al Thornton, Ronny Turiaf, Kelly Olynyk, and Matthew Dellavedova. In 2014, with the SIMI motion-capture system in place, P3 would begin a new relationship with the NBA, where they'd assess almost every player at the NBA draft combine, which delivered an explosive new dataset.

Players began visiting Santa Barbara all the time. In the early days, Eric got a little starstruck. When NBA players came in, Marcus says, "Eric's knees used to get weak. He would have trouble talking to people. You know, 'Vince Carter! I can't fucking believe this!'"

A practical problem with Eric's shyness was that, typically, a few minutes after meeting an athlete like Carter, Eric had to convince him to strip so that Eric could place twenty-two markers all over. P3 buys B&L brand 12.7 millimeter pearl markers on a flexible three-quarter-inch black naugahyde base. Six dollars each, they feature a knot of reflective fabric the size of a pencil-tip eraser. The expensive cameras of the motion-capture system can tell a ton about where those are in space—so long as no flap of fabric blocks the view. P3 needs people in the form-fitting tights that athletes tend to wear as a base layer—and nothing more. (Now P3 has spandex shorts, jog bras, and tights for borrowing.) Eric has asked many of earth's finest athletes to take almost all their clothes off in full view of everyone else in the gym.

Eric found this "terrifying."

"No words. You put a few on the ankles, a few on the knees. And at that point, once you've done that, like, the other ones are so far down there, right there." Eric indicates either side of the privates. The spots have names like "perineum bone left" and "pelvis."

"At that point, I just say, without even looking—certainly without looking them in the eye, just staring off into space—'Gonna throw a few on your hips and on your back and off we go.'"

The first step was to understand the different events in sequence. In biomechanics textbooks, a jump is taught as having a force-time curve with three turns. There's a low point of force while unweighting, then a turn streaking up to a peak just before leaving the ground, and then curving back to zero once the jumper is airborne.

But Eric's data looked different. Some force-time curves had big initial impact spikes, some displayed hard-to-fathom wiggles. Many of the best jumpers in the NBA population had six points on their force-time curves. Just for fun, Eric figured out how to become a six-point jumper, but since the data was inconclusive, Eric refuses to speculate on whether this was worth doing.

Despite displaying clear patterns, P3's early motion-capture dataset produced no solid conclusions. The data itself could overwhelm. The cameras shoot 120 frames per second. The force plate works at 1,000 hertz. They recorded nearly a million data points per assessment. Eric says his team would "pull out more discrete variables, about 500 data points per athlete per assessment. For the one movement of the drop jump, that's probably about 150 or 200 columns of data, give or take."

As the rows and columns filled, however, Eric noticed something intriguing. One of biomechanics' favorite factors is something called the *Q angle*. That's the catchy name for the quadriceps angle from your hips to your knees—a data point that has been obsessively studied for decades.

Not long after World War II, a researcher named Håkan Brattström wrote letters to people all over Scandinavia who had suffered knee dislocations, hoping to put them in his Danatom Type A X-ray machine. He spent a decade recording Q angles, which he summarized in a classic 149-page study concluding that bad Q angles affect the kneecap.

In the years since, dozens of journal articles have affirmed that knees like Kyle Korver's, which collapse toward each other while bending (also called *valgus collapse*), are at high risk for injury. Valgus collapse is to athletic movement as drunk driving is to automobile safety.

Ramada R. Khasawneh summarized the field's findings in *PLOS One* in 2019, saying "it is beyond doubt that misalignment will cause problems to the knee function. Therefore, the determination of the Q angle is particularly momentous for patients who are athletically and physically active."

Bad Q angles also seem to address one of the biggest mysteries of sports injuries: Why do boys and girls have similar rates of ACL injury in grade school, but then in the teenage years the girls get hurt far more? A tidy theory is that in puberty, girls' hips generally grow wider, which, geometrically, creates space for bad Q angles.

The twenty-two reflective points that Eric personally arrayed around bodies did a perfect job of measuring Q angles—what Brattström would have given for data like that! Arrayed in three dimensions, those twenty-two points create 231 angles. One of them is the right-side Q angle, and another is the left.

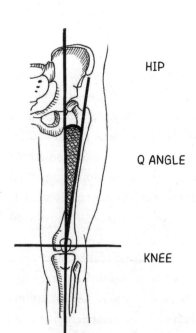

HIP

Q ANGLE

KNEE

The problem was that the columns for Q angles showed no relationship with catastrophic knee injuries. Eric started to wonder if the link between Q angles and ACL tears had been affected by what psychologists call *availability bias*. Q angles were easy to see. Everyone had long thought Q angles mattered. That alone made them well studied, and much discussed.

P3 posted Instagram video of Scottie Barnes jumping over tiny hurdles. One of the first comments was from a doctor of physical therapy in New York, who suggested they ought to address his Q angles before he tore an ACL. But by the time Barnes bounced over those hurdles, P3 was eight years into collecting detailed movement data from large numbers of athletes and correlating that with outcomes. "We don't get as concerned about that kind of classic knee cave-in as we used to," replied P3 head trainer Jon Flake.

Other research has begun to find similarly: in one study, female soccer players who tore their ACLs were found to have Q angles that were no different than players with intact ACLs. Another study found similar results in men, while a third found no difference in Q angles in MRIs of those who did and didn't have ligament tears.

This is not to say that valgus collapse is meaningless. For one thing, that collapsing action, Jon says, "can put you at a higher risk for other minor knee injuries or a kind of long-term wear and tear on the patella. But in terms of catastrophic ACL tear . . . it's not as concerning as maybe we once thought." Moles *can* signal cancer. But sometimes a mole is just a mole, and in P3 data, sometimes valgus is just valgus.

P3's early movement data mostly made the case that knees—and all other joints—are complex. It's seldom the case that one simple, visible factor like a Q angle is the skeleton key.

Weight is another example. As an isolated factor studied in academic journals, extra mass seemed not to affect injury risk. "Using these single variate analyses, it doesn't appear to be a risk factor at all. It's not even close to statistical significance," says Marcus.

But think about the destructive force of Katie Spieler's heels hitting the ground. You only need high school physics to know that an extra twenty pounds would put measurably bigger forces into her knee.

And that's what P3 found: extra weight was a force multiplier for those at biomechanical risk. Being heavy makes bad landings more dangerous. "When we use this machine learning model," Marcus says, "weight becomes the third most important variable in traumatic knee injuries in NBA players—because it's a modifier of these other risk factors."

"You can be three hundred pounds," says Eric, "and land just fine—is what the numbers would suggest. But if one or two of those bolts [he means good landing technique] starts to come undone, and you weigh that much, that risk is going to balloon pretty quickly."

The force of landing travels from the floor up through the body, and it can do that by many different paths. If your joints are nicely aligned, those forces travel on the body's superhighway, through big, strong soft tissues, which evolved to carry big loads. Semitrailers don't damage interstates much, but they shred back roads in small towns, which in the body are the tendons and ligaments of the leg joints.

Eric's challenge was to map the back roads. Two-thirds of NBA players bent all six of the big joints of their two combined legs with each drop jump. Ankles, knees, and hips all gave at once, dispersing newtons across calves, quads, and glutes. Marcus and Eric started calling those players "yielders."

But what was happening with the other third of NBA players? Some of them could jump crazy high. But while yielders had force-time curves lined up like sweaters at Nordstrom, the other players' data looked like a teenager's bedroom floor—some flat bits, here and there a knobby hill, and then perhaps a worrisome Matterhorn. "Instead of seeing six discrete time points, you'd see four. Or three, and some emphasis," says Eric.

Untangling that "was not a fun part of the story," explains Eric. It required "looking at hundreds and hundreds" of data points and comparing them with their better-moving colleagues. Even naming this second group was hard. Marcus proposed calling them "non-yielders." But what they did wasn't the same from player to player. They did a mix of things. Or a blend. Eric named them "blenders."

Did blenders' feet hit the ground differently? Did their ankles roll to the side? What about asymmetry? Did their feet rotate? Did their knees? Every theory could be checked; the checking devoured Eric's schedule. P3's tiny kitchenette has a sink that fills with plastic tumblers from athletes' protein shakes. Espresso-stained cups, from Eric and his growing data team, joined the mix.

Marcus paid obscene bills—for the SIMI, for the servers, for the coffee. And mostly, for the outrageous chunks of time spent processing every assessment. And he knew the dumbest approach would be to rush to conclusions. Datasets must breathe and grow. The early findings, from scant data, can look silly in time.

No one remembers the exact day Eric solved the blender riddle: Hips! It was the hips. Blenders had ankles and knees that worked the same as yielders, but their hips never pushed back and low as they landed. Blenders loaded calves and quads on landing, but not glutes. Blenders omitted the biggest, strongest, most useful joint of the bunch.

The confusing part was that blenders avoided their hips in different ways. Some hips did basically nothing. Others hips tried, but were too rigid to sink meaningfully into a landing. Another eager group had the timing off: as the feet hit the floor, and the knees and ankles relaxed into eccentric loading, the hips were already into the next step, starting the jump. When a normal yielder loads, the hips move backward as the body sinks down; Eric watched those blenders on video and saw the shiny pearl markers on the hips move *forward*. It's like a drummer counting in the next song while the band finishes the last one.

Sometimes Eric gets a little nervous about presenting findings in staff meetings; P3 employs intelligent, scrutinous people who challenge assumptions and ask hard questions. But on the day he presented his findings on blenders, he knew he had something good: *among NBA players, blenders are 300 percent more likely than yielders to have back trouble.*

The implication was that much of the NBA, medical science, and the rest of us, too, had missed one of the big causes of lower-back pain. The most common advice to fix back pain is to stretch and strengthen the core. Around the league, players with sore backs spent hours doing that work under the watchful eyes of physical therapists and trainers. But in many of those cases, Marcus says, "that has nothing to do with it. Their cores are perfectly strong and flexible."

Instead, blenders land without the helpful control of the big muscles of the hips and posterior chain. When the hips don't shift the force of landing to the glutes, impact forces travel up into the lower back, where they produce a terrifying sounding effect called *anterior shear*. Anterior lumbar shear is associated with an array of lumbar damage—disc compression, spondylosis, capsule tears, endplate compressions.

Alarmingly, many different researchers have applied anterior shear forces to dead bodies to see how hard it is to break a human back. A 2012 *Clinical Biomechanics* article rounds up the findings and reports "cadaver anterior lumbar failure started at 1200 N[ewtons] and hard tissue failure occurred at the 2800 N level." The paper, which is mostly focused on jobs involving repeated lifting, recommends a seven-hundred-newton limit for workers who cope with repetitive forces. "The ultimate shear strength of human lumbar specimens," write Sean Gallagher and William S. Marras, "ranges from 0.6 to 3.2 kN."

P3 measured lean, five-foot, five-inch volleyball player Katie Spieler landing on the floor with 3,100 newtons, and once measured a basketball player with a peak force of 11,000 newtons. Clearly, the back is counting on glutes and other body parts taking most or all of that force, or else it would rip itself apart.

Not long after Eric discovered the connection between blending and back pain, most of the P3 staff was scheduled to head to Chicago to assess the 2014 NBA draft class, which featured players like Andrew Wiggins, Joel Embiid, Julius Randle, Marcus Smart, and Zach LaVine. This was a completely different population than had been assessed previously, and a chance to play Eric's favorite game: replicate the findings.

That's also when Eric stopped being shy about asking players to strip. His coworkers make fun of the catchphrase that emerged, which he still says just about every time: "I can see if they have tights on already," Eric says. "If they do, then I say, 'Shorts off, let's do this.' And that's just kind of how it goes."

The results of the second study offered near-perfect confirmation. "The numbers were almost exactly the same. A third of the athletes were blenders. And those blenders had about a 300 percent higher incidence of back injury on their timeline history. Those are super-compelling moments. Oh my God. The clinical data checks out!" Marcus says. "Your back is going to have to absorb a lot more shear forces, or that force has to go other places if it doesn't go through the hips."

With Eric's findings rolling in, Marcus started to really dig motion capture. He'd spent his whole life looking for evidence of injuries before they occur, and here they'd found "a ridiculously strong signal."

Eventually, Eric even returned to his initial project: understanding jump height. It turned out that the prescription for a good jump was like so many prescriptions: different by patient. That is to say, that P3 confirmed the findings of other studies, that the most important factors in determining jump height—for all players—was how hard you can push on the floor, how quickly you can straighten your bent knee (acceleration and peak speed both matter), and height—shorter NBA players have higher verticals.

But prior research into NBA players suffered from small sample size (one prominent study assessed just eleven athletes). P3 published research on the jumping strategies of 178 players. What they found was that it was wrong to recommend any strategy to every player. Instead, the analysis identified three clusters, each with their own movement characteristics and recommendations.

The group called *stiff flexors* jumped without bending any joints much. They put huge forces into the ground by being quick, carrying momentum from the run-up into jumping higher. You might expect that P3 would want to get those players moving through wider ranges of motion in the knee and ankle, but interestingly, additional movement didn't help this group of players jump higher. So, for them, P3 recommended quick and snappy half squats, which would let them develop more force without implementing deep knee bends that would make them slower on the ground.

A second group, *hyper flexors*, bent significantly through the ankles, knees, and hips. That takes time; they were slower on the ground, and generated smaller forces—but jumped just as high as stiff flexors, thanks to the magic of the body's built-in rubber bands. For them, P3 recommended generalized strength work, a wide array of weighted exercises, to make their rubber bands stronger and snappier.

A third group, mostly among the league's tallest players, were stiff in the ankles and knees, but bent so much at the hip that their chests would often end up parallel to the floor. P3 called them *hip flexors*. Hip flexors had an approach that relied heavily on glute tissue—for them, P3 recommended aggressive strength training in the posterior chain to give them increased force without having to change their mechanics.

Meanwhile, P3 held internal meetings about how to steer blenders away from the risks of back injury. The coaches and biomechanists all guessed the move would be to train the posterior chain. For blenders, Eric says they combined traditional strengthening like Romanian dead lifts with an array of glute exercises, as well as "lots of coaches yelling at athletes" during plyometrics, drilling the technique of getting hips back while landing. "That approach," Eric says, "has withstood the test of time."

Halfway through a long bike ride, on my second trip to Santa Barbara, Marcus said something in passing that stopped me cold: "You've got the look of a blender."

# 10

## GRAVITY

*You can't move mountains by whispering at them.*

—Pink

Santa Barbara has boulders. Lining Rattlesnake Canyon, piled on the beach grass, and along the Elliotts' driveway. "If you ask people how the rocks got there," John McPhee once wrote, "they assume it was by a process that is no longer functioning. If you suggest that the rocks may have come from the mountains, people say, 'No way.'"

In a 1988 *New Yorker* story, McPhee explains: tectonic plates collide, nudging Southern California's mountains taller but also destabilizing and crumbling the sandstone. Wildfires incinerate the plants whose roots hold it all together, and change the soil structure so it repels water. Then the rain comes.

McPhee says government engineers visited a hamlet called Hidden Springs. They explained that 400,000 cubic yards of rock, mud, sand, and debris lingered just uphill and the next good rain could take out the whole town. Thirteen died in the next rainstorm as boulders were deposited miles away.

Nature, it seemed to many, was as wild and unpredictable as a heart attack. "The fact that people did not believe what could happen was disappointing, actually," one of the engineers told McPhee.

When Derrick Rose joined the NBA, he thumbed his nose at the notion of human frailty, attacking the rim with dangerous ferocity, like a thicker, springier Allen Iverson.

Physics bursts with forces. But highlights, Eric points out, are always the same force: acceleration. At that, Rose was legendary. He pushed off the floor and led *SportsCenter* with two-handed hammer dunks, twisting reverses, and scoop shots. Many highlights ended with Rose floating, Jordanesque, as gravity plucked defenders from the air around him.

Alas, physics' less showy forces also applied to Rose. There were collisions of all kinds, but especially between Rose's size 13 Adidas and the squeaky-clean floorboards of the National Basketball Association. The copious newtons of D-Rose's sharp landings reverberated through his body. He became the NBA's youngest-ever MVP at age twenty-two, and then lost more than half of the following season to turf toe, groin strain, back spasms, and sore feet.

That year's playoffs opened with an afternoon visit from the Philadelphia 76ers. The Bulls were up a dozen with a little over a minute left, when Rose—inexplicably still in the game—ditched his man near half-court.

Three quick dribbles later, 76ers big man Spencer Hawes was the sole remaining line of defense. This is how highlights are born. Rose started left, veered right, gathered the ball, and pitched both feet at the floor like he was trying to crack a frozen puddle. His left heel struck with a leg that was straight until the knee buckled. Rose bounced into the air and shoveled the ball to a teammate from an airborne fetal position. By the time his teammate Kyle Korver called the game "the saddest win," Rose was already at the hospital.

Rose rehabbed his torn left ACL at a Los Angeles facility called Athletes' Performance, under the watchful eye of Rose's two employers: the Bulls and Adidas.

Eighteen months later, Rose was, they said, in incredible athletic condition, strong as a Bull, and—the reports were breathless—*jumping five inches higher.* Rose arrived at training camp in 2013 looking like an NFL player.

"It was such a moment," says Marcus. The head of Adidas basketball was at P3 in Santa Barbara as news emerged of Rose's insane vertical. "I told him it doesn't matter if he jumps higher. . . . What was important was that he didn't have new compensations that created more risk and broke something else."

It's hard to know when a recovering player is ready. But P3 insists that you can't tell whether a knee is good to go by looking only at the knee. Bodies move differently over time—they *compensate*—especially after trauma like a torn ACL, surgery, months of rehab, and, in Rose's case, adding a medium-sized dog's worth of new muscle. Marcus says compensations haven't been well studied but appear to be a giant factor in sports injuries.

In the tenth game of Rose's return, he cut along the left baseline, trailed by Portland Trail Blazer Nicolas Batum. Joakim Noah slipped a lovely two-handed pass that might have led to a Rose dunk. But Batum stuck out an absurdly long arm and touched the pass, which ignited—surprise!—a one-yard race in a new direction. With expressions from Renaissance paintings, Rose and Batum lurched after the ball.

That's when Rose's "good" right knee buckled inward. In real time, it was nothing—a weird step. In retrospect, it was the end of Rose as a player for the season and, at age twenty-five, as an MVP candidate forever. He had torn the meniscus in his right knee, which ultimately proved to be even more troublesome than the left.

Before Rose played in another NBA game, P3 got a contract with the league to assess draft candidates en masse. No one knew what to make of the biomechanists at the 2014 draft combine. When the P3 crew flew back to Santa Barbara and processed all the data, Marcus says they identified giant red flags of injury risk in many top players. No one knew what to make of that, either. Or, rather, no one knew *to* make anything of it. At the time, P3 sent all of the information only to the league.

Eight of the top eleven draft picks—largely, Marcus says, from P3's list of red flags—suffered major injuries almost immediately. Second-overall pick Jabari Parker missed fifty-seven games. Third pick Joel

Embiid missed his first two seasons and most of a third. The Los Angeles Lakers' great hope, Julius Randle, ended his season on a routine-looking drive to his right thirteen minutes and thirty-four seconds into his career.

ESPN's Baxter Holmes published a two-part investigative series that quoted surgeons, team trainers, coaches, players, and parents expressing alarm at the epidemic of injuries in young basketball players. Marcus is quoted telling Holmes that a lot of today's elite athletes peak at sixteen or seventeen, and then begin succumbing to wear and tear. The commissioner of the NBA, Adam Silver, tells Holmes that getting young players to the league healthier is "the highest priority for the league."

Marcus felt deep regret when many of those players fell to injuries that the data saw coming. After that year, Marcus says P3 began sending the assessment data to the players directly—so that it has a chance of being helpful.

Around that same time UCLA's Zach LaVine appeared at P3 for pre-draft training. Eric says he had never seen an athlete like that nineteen-year-old. Years later, he made a chart of all the NBA players P3 has ever tested, plotting them by lateral athletic ability on the X axis and vertical athletic ability on the Y. LaVine exists in the far top-right corner "with," according to Marcus, "Anthony Edwards and God's chosen children."

The NBA says it features the best athletes in the world, and it might—but Eric points out that it's hard to say for certain because NBA players have tuned their bodies for skills that matter most in basketball, rather than, say, the 100-meter dash or the decathlon. But Eric says Anthony Edwards would have a good shot at winning gold in something besides basketball and he wonders if LaVine, at his peak, might have won an Olympic high jump.

LaVine's 2014 assessment showed some concerns in his left knee, mostly stemming from his left foot striking the ground with a lot of force. But to the naked eye, his performance was mind-blowing. The Minnesota Timberwolves drafted LaVine thirteenth overall, he put on a stunning show of dunks in warm-ups for his first summer league game, and he flew around the court, ferocious and nimble, barely missing any games that season.

LaVine returned to P3 the following offseason carrying the bio-mechanical residue of seventy-seven intense competitions. An internal P3 document says that, as of his May 23, 2015, assessment, LaVine had "clear mechanical risks," especially to his left knee. In February 2016, LaVine won one of the most incredible dunk contests of all time against another P3 client, Aaron Gordon. A year after that, in February 2017, LaVine rose to score over and around Andre Drummond, landed hard on his left heel, and created a moment of violence in his knee. He played a bit more. But after the game, LaVine needed his mom to calm him down after learning he had a torn ACL. The Timberwolves traded LaVine to the Bulls and he began grueling rehab.

In the autumn of 2018, Marcus got chills, a haunting déjà vu, when he heard that an explosive Bulls guard with a torn left ACL had returned from rehab jumping five inches higher. "This," says Marcus, "was an exact replay of what happened to D-Rose."

Zach had crushed all of the Bulls' performance testing... *but so had Derrick Rose.* "Traditionally," says Marcus, "the primary metric for return to play for ACL was time, six to eight months, and enough strength—primarily extension strength in the injured knee." The standard goal, Marcus says, was to get the rehabbed knee within 90 percent of the strength of the healthy leg.

"Think how complex and precise the timing and coordination are to play this big man's ballet," says Marcus. "It's very easy to get strong and still move very badly, or to lose strength in both legs, allowing you to be at 90 percent of a weak leg.

"Really good athletes tend to be compensation savants. They can change mechanics, and performance may not fall off at all. They can run really fast, jump really high, and look amazing—even though they are doing things in ways they were not designed, and that could be career threatening."

People don't hesitate to accept science when it's obvious. People don't often argue with "you have torn your ACL." But the more useful "you're at high risk right now" inspires the same skepticism as engineers saying your town's going to be swallowed by mud in the next rainstorm. A scan of LaVine's knee eight months after surgery would have shown healthy tissue, just as every building in Hidden Springs was intact while the engineers warned of the landslide.

Marcus decided to make a fuss. He didn't want a second helping of regret. He called the Bulls and suggested that they send Zach back to Santa Barbara.

The Bulls agreed, and the assessment found some good news. Through his postsurgical recovery, the Bulls staff had been consciously working on LaVine's hip strength, and one of his numbers had improved, impressively, from among the very worst in the NBA to the 24th percentile.

But there was scary news in LaVine's ground contact numbers on both sides. LaVine didn't just land hard, from great heights, but the forces of the floor explored some of the iffier back roads as they traveled up his body. LaVine had long landed with floppy-ankled foot positioning that put his knees at risk. He'd never had ground contact numbers that P3 saw as good; now he had some of the worst in the NBA.

"The fact that Zach was jumping breathtakingly high," says Marcus, "clouded the fact that he was not close to ready and was at much higher risk mechanically with a bunch of new compensations." Before he played in the NBA, P3 had assessed risk in Zach's left knee, and it had since endured a serious injury. Now they assessed giant risk in his right knee, ankle, and foot, too. It wasn't a sign that anything had grown weak in Zach, but rather that, years into a grueling NBA schedule, and months after an intense injury, his movement patterns had shifted a bit. It's common, after a major injury, to shift weight to the other leg.

On a conference call with the Bulls, Marcus suggested aggressive ground-foot interaction training—a mix of ankle hops, pogos, and skaters, with feedback throughout on positioning his foot during landing. LaVine's case also called for improving ankle stiffness with calf raises and isometric calf holds. As his ground contacts improved, P3 recommended that LaVine progress to bigger jumps on and off boxes and platforms, laterally, off one or two legs.

P3 also recommended that the Bulls continue the hip stability work they had been doing, and add a wrinkle. "He didn't absorb much force through his hips," says Marcus. They recommended dead lifts and other posterior chain moves, and mobility work designed to improve his internal hip rotation, which would help him absorb forces of landing and cutting.

LaVine flew back to Chicago and "worked specifically on the issues that the data showed had changed for the worse or were already dangerous," says Marcus. That last sentence should, in Marcus's view, be common, but he says it with delight and wonder. Here was a case of people believing the engineers and avoiding the landslide.

As LaVine finished two months of pogoing around the Bulls' practice facility, one of the biggest wildfires in California history began in Ventura—between Los Angeles and Santa Barbara. The Thomas fire burned more than 250,000 acres and a thousand buildings. It spread in many directions, including into the steep hills above Santa Barbara.

There was smoke in the air when LaVine flew back to Santa Barbara for his follow-up assessment in December 2017. His key ground contact numbers had returned to their pre-injury levels—from the 2nd percentile back to 9th. Almost every other number up the chain of his joints had joined the NBA mainstream: a knee-stress measure had climbed from a scary 3rd percentile to 38th, while a hip measure that had been as low as the 8th percentile was now 48th. Marcus saw it as a triumph.

P3 is in the business of creating nonevents. A nonevent for Zach LaVine meant another brilliant year of basketball.

About that time, Marcus went for a run above town and came back telling his family the soil didn't feel right. It was spongy. "Everything people on the East Coast imagine will happen to California," Marcus told me by phone in the first week of January, "is happening here right now." I was planning my first visit.

Then it began to rain. In the predawn of Tuesday, January 9, mudslides collected dogs, cars, trees, houses, and families. Many had ignored evacuation orders. The muck carried twenty-three to their death. Dormers, boilers, books—a gruesome salad spewed across school grounds, up the overflowed banks of streams, and across Highway 101. The crash course in boulder travel had shut off Santa Barbara from the rescuers of Los Angeles.

On Thursday, Marcus told me to come anyway—we'd figure it out.

In Ventura, where the fire began, I joined an eager queue waiting to board a whale-watching boat that had been pressed into temporary service as a ferry. As we waited, an exhausted-looking woman in tattered pants fussed over a small dog while her husband worked his phone, trying to find out who among his neighbors, friends, and family had survived. Even before docking, you could see firefighters walking in lines, poking the muck with long sticks.

I walked to P3, met Marcus and Eric, and quickly began learning how our bodies have big forces stored uphill like debris above Hidden Springs. Eric and Marcus sounded like the engineers in McPhee's story, with a clear view of what might muck up our futures.

One of their stories is about a high NBA draft pick who showed up with his team in the days before his first NBA game. His assessment mostly went well. He jumped beautifully. His movement was NBA-grade. He had the tools.

But the force plates saw something. "When he would land from a jump, for a second jump, he would always keep the left heel off the ground. It would never touch the ground," Marcus says. "It's very, very rare that we have an athlete that lands from a jump and goes up for a second jump, and their heels don't touch the ground. It never happens—unless there's compensation for a big, big injury—or a big mechanical issue."

The left heel's reluctance shifted the force of landing to the right. "Significantly more force through his right side on every landing," says Marcus. "Over 60 percent of his net force is being absorbed by his right side."

P3's motion-capture system—shiny and new at the time—saw a thicket of risks to the right knee. Not only had his weird movement added force, but it had also complicated the player's biomechanics. ACL tears are rare, but they're much more common in knees that move like that one.

P3 doesn't have a private exam room or a white noise machine to deliver news in privacy, but when necessary they huddle close and share secrets. The team honchos, across the room on the couches, wouldn't hear a word over the music.

Marcus and Eric told the player they saw worrisome forces through his right knee. "That resonated," Marcus says. "He told us that his right knee *had* been hurting. And then he told us that he had been shut down for the last month of the summer, meaning he wasn't playing basketball because he was hurting that much."

Baseball and football players go weeks without playing their sports. But ballers ball. Almost no matter what P3 tells NBA players to do, they play. Basketball has pull. Sitting out a month of this game at that age set off alarm bells.

"And then," Marcus says, "he told us he got an MRI, off the record, two weeks ago." Only his agent knew. Thankfully, the MRI showed no major damage—but he was still in pain. "To have great pain, is to have certainty," writes Elaine Scarry in *The Body in Pain*. "To hear that another person has pain is to have doubt." In other words, pain is invisible.

But P3's technology makes pain somewhat visible. Nobody had noticed the player's elevated heel, or sore knee, during a well-televised season. But the SIMI saw them. "He starts revealing more and more and more because he has confidence that we have information that he needs," says Marcus. "We're predicting the future. And he's saying, 'Well, the future is here right now.' "

Training camp began in a few days. It would mean in-your-face physical tests against the best in the world. Playing time, scoring opportunities, and ultimately millions of dollars were at stake. The player said he planned to play.

That may sound silly or brutish. But it's common. Pain is the invisible lifeblood of elite sports. (What's the reason most of us don't know how fast we could really run? Because it would hurt, right?) At the bleeding edge of performance, the limits are regularly expressed in suffering. Tour de France racers describe their own chosen sport as a "suffer fest." One has a story of grinding his teeth flat as a more-pleasant distraction from the race's other pain. The NBA lauds Isiah Thomas or Willis Reed for playing badly hurt. Professional athletes touch those limits for a living. This player had a clean MRI, whippersnapper youth, and elite athleticism; to sit would ruin his reputation.

The fact that the player had tested as one of the best movers on the planet on one and a half feet is mind-blowing. His body invented a new movement pattern on the fly, and it worked. He could *move*.

"But now the problem is it's threatening his career," Marcus says.

"After he told us all this about his knee, he told us that he had had plantar fasciitis on his left foot," says Marcus. "So, this entire time when he's going out trying to be the best player in college basketball, his left foot hurts. He developed these movement patterns that saved his left foot."

And it worked. "His left foot," says Marcus, "is perfect. He doesn't have any scar tissue, the tissue quality is perfect, and mobility is perfect."

The fasciitis was cured. But he still won't put his left heel down. Because "pain is an amazing teacher," says Rachel Zoffness, Ph.D., a pain psychologist, Stanford lecturer, and assistant clinical professor at the University of California, San Francisco School of Medicine. The human brain might be the most beautiful and intricate of all creations. But it makes mistakes. Pain, according to Zoffness, is "the brain's opinion" about the danger you're in.

Zoffness tells a story first published in the *British Medical Journal* about a construction worker who jumped off a ledge and—oops!—onto a seven-inch nail. The nail poked out the top of his left boot and popped the balloon of normalcy. At the ER, they administered intravenous fentanyl and midazolam before very carefully sawing away his boot.

Inside, though, they found a perfect left foot. No blood, no wound, no nothing. Surprise! The nail had threaded between toes. The worker's brain had formed the opinion that the body was damaged, and sent pain signals—not because it felt the nail tearing flesh, but because it saw a nail poking through a boot top.

The brain can project real pain into pristine flesh. It can even project real pain into no flesh at all. Amputees can experience pain, called *phantom limb pain*, in an absent arm or leg. The homunculus is the brain's map of the body. One way to treat phantom limb pain is *mirror therapy*, in which the patient looks in a mirror while completing prescribed movements. Eventually, the homunculus updates with better information about where the body ends.

"The word for what's happening is *kinesiophobia*," says Zoffness when she hears about the rookie who wouldn't put his foot down. In the literature, kinesiophobia is defined as "an excessive, irrational, and debilitating fear of physical movement and activity resulting from a feeling of vulnerability due to painful injury or reinjury." It's common, even in healthy tissue. Sometimes the brain's opinion of danger is wrong.

"You're terrified it'll hurt again," says Zoffness. That construction worker's ability to use his left foot improved dramatically once he would see it wasn't damaged.

"The solution is that his brain needs to learn that it's safe to put his whole heel down, but the brain won't know that it's safe without concrete evidence," says Zoffness.

In the back of P3, they devised a program. But mostly, they cheered him on when he got his weight distributed evenly. The team helped coach it, and after a few weeks he had freed his movement habits from pain's hangover. "And that was all it took," says Marcus. "It wasn't building new systems. It wasn't rehabbing anything—nothing was broken. It was just giving him feedback on when he was more symmetric, and when he was getting his heel down to the ground."

P3 wanted to give Zach LaVine better movement habits, and the confidence to fully rely on his repaired left leg. Free movement is healthy movement. It's hard to pull that off unless you feel safe.

Before the mudslides, the plan had been to meet the Elliotts at a restaurant. Nadine asked if we could pivot to comfort food at home, where she stood over an enormous pot of lentil soup as the children made inventive paper airplanes. Marcus fetched some kind of big-deal wine bottle from his stash in the garage. (Marcus, Nadine says, "is becoming quite a little wine snob!")

Most of dinner was spent leaning in to hear the aftermath. Nadine had updates about how the mudslides had affected families of the children's classmates. A fire flickered, fending off the cool evening air tumbling through the cracked patio door.

Before long, Kira had loaded the dishwasher and disappeared into her room. Nadine was reminding Keean about a straggler homework assignment, while the youngest, Mila, made progress on her mission to watch every single episode of *The Simpsons* twice.

With a half-bottle of snob wine and two glasses left on the table, Marcus and I lapsed into a conversation about misperception. I had just met with a neurology professor in New York, who had set me off on a reading tour about how we contort the truth—going back to the days of Plato's cave. Hannah Arendt wrote in the 1960s that she believes Homer *invented* truth. It's hardly surprising, then, if in today's sports world we have a hard time agreeing on, say, what causes injuries.

In fact, sports played a starring role in the research. A paper called "They Saw a Game" is a masterful examination of selective perception. The researchers examined a 1951 Princeton–Dartmouth football game, and noted that rooting for Princeton—whose star quarterback left the game in the second quarter with a broken nose and concussion—made you see things differently than rooting for Dartmouth. They surveyed fans of both teams over time, showing the same footage. "Results indicate that the 'game' was actually many different games and that each version of the events that transpired was just as 'real' to a particular person as other versions were to other people," write Albert Hastorf and Hadley Cantril.

The study concludes that "it is inaccurate and misleading to say that different people have different 'attitudes' concerning the same 'thing.' For the 'thing' simply is not the same for different people whether the 'thing' is a football game, a presidential candidate, Communism, or spinach."

NYU's Jay Van Bavel, a researcher of psychology and neuroscience, explained that, roughly speaking, our brains have two systems of thinking—in keeping with Daniel Kahneman's Nobel-winning *Thinking, Fast and Slow*. One is quick, emotional, and right at home in watching and rooting for sports teams. The other is slow, rational, and the key to sorting truth from fiction. I studied the training manual for CIA intelligence analysts, whose jobs depend on sober judgment. It's an exercise in removing every last shred of emotion—of course football fans failed that test. Rooting interests obliterate accurate perception.

The Bulls had embraced insight, borne of SIMI data, about Zach

LaVine's vulnerable landing patterns, and held him out of games. Those kinds of evidence-based decisions might be common in medical settings, built around evidence and the search for truth. But in Marcus's experience they're rare in sports, which might be the central frustration of Marcus's career. Teams have been glacially slow in adjusting to the evidence. Many of the best athletes in the world have detailed movement data on Eric's servers that teams discard or ignore. Sometimes it costs careers.

The wine was Nebbiolo—a grape whose name means "fog" in Italian, because it's harvested in October's gloomy mist. As I was suggesting that truth is tricky to discern, Marcus lifted his glass—not to eye level as if to toast, but higher, arm lifted skyward from a tall torso with upright posture and a full range of shoulder motion. It felt ten feet off the table.

I stopped talking.

"It's Newtonian physics!" Marcus looked me dead in the eye. "If I drop this glass, we all know what's going to happen. It's going to fall; it's probably going to break." The force that tears an ACL is the same one that carries boulders to the beach and spills Nebbiolo. Physics doesn't care if you believe it or not.

After lunch the next day (and an apology for his impassioned speech the night before), Marcus drove me to catch the ferry, which pulled out past a yacht called *The Truth*. That night, LaVine returned to the court. A few minutes in, he drained a three. He wove through traffic and raised to dunk with ferocity. LaVine went years without missing any big chunks of time since his ACL surgery, and has been an All-Star twice. "Fortunately," says Marcus, "it worked beautifully."

# 11

GROUND UP

*I will land on my feet with a smile on my face.*

—Diane Cook

THE MATH WAS SIMPLE, once you knew how much they weighed and how far they flew. In the late 1950s, Soviet track coach Yuri Verkhoshansky calculated that his triple jumpers pushed on the earth with about three hundred kilograms of force—nearly three thousand newtons.

And yet, not one of them could squat anything close to that much. Evidently, athletes found some extra juice while jumping. Where did it come from?

This was a mystery Verkhoshansky wanted to solve. Perhaps shorter motions permitted bigger forces? Verkhoshansky led a session of half-squat workouts, encouraging his athletes to attempt wildly heavy weights. Sure enough, his jumpers could handle much more—but still not three hundred kilograms. And it seemed like a bad sign when none of the jumpers made it into the gym the following day, because they were all in agony with back pain. For a while, Verkhoshansky had athletes lie down, and lift the bar with coaches holding the bar atop their feet.

Eventually, Verkhoshansky realized that the secret ingredient that made athletes so much more powerful while jumping was speed. Triple jumpers touch down for just a tenth of a second. "The colossal load that acts on the athlete's leg," Verkhoshansky wrote in his book *The Shock Method*, "is the kinetic energy accumulated by the athlete's body during the run-up. In that moment a new idea came to me: why not try to use a training exercise in which the kinetic energy of the falling body will be applied as the external load instead of a barbell?" That's how Verkhoshansky pioneered training human ballistics, starting with depth jumps and progressing into many explosive movements. It was a sports science breakthrough.

Soon the Soviets dominated Olympic medals. In 1975, American runner Fred Wilt noticed that Soviet runners would warm up with all kinds of jumps. Meanwhile, the Americans stood, stretched, wondered what the Soviets were up to—and often lost. Wilt (who was also, no joke, an FBI agent) investigated, coined the term *plyometrics*, and helped to popularize Verkhoshansky's methods around the globe.

Marcus found much to quibble with in the Soviet system—for starters, the rampant doping and drudgery. But there was no denying the findings of the world's first real scientific dive into sports performance. Marcus ingested everything Verkhoshansky wrote, and baked big, aggressive plyometrics into almost every training program.

In the late 1990s, after Harvard but before P3, Marcus made recurring visits to coach some of the best track and field athletes in the world alongside Canadian coach Brent McFarlane. They talked constantly about how bodies handle the kinds of forces Verkhoshansky wrote about. To athletic trainers, weight-lifting sessions long seemed like the moment the body was most stressed—which is why squats and cleans come thick in orthodoxy about safe best practices. (Where there are barbells, there are often people to remind you to keep your chest up and your hips back.) Meanwhile, explosive jumps and landings feature far greater forces—but athletes were expected to take off and land unsupervised.

The topic dominated many lunches for Marcus and Brent. You could succeed in soccer with elite ball skills; basketball players might hang around thanks to their ability to shoot or read the floor. The hurdlers, sprinters, and long jumpers, though, could only reach the highest levels by training their bodies to safely and quickly control the biggest forces bodies ever encounter. "Their livelihoods are determined," Marcus explains, "by not giving up hundredths of a second anywhere." And while basketball and football players touched their feet down in myriad ways (toes down, heel down, you name it), in elite track and field, almost every foot touched the ground dorsiflexed.

In strict biomechanical terms, *plantar flexion* means using your ankle to point your toes at the ground like a ballerina *en pointe*, while dorsiflexion means the opposite: pulling your toes up toward your shin. In practical coaching terms, Marcus found it easier to tell people to land "toes up." They would seldom land with their toes actually high off the ground, but saying that would nudge them closer to Marcus's true goal of hitting the ground in one clean motion, with the ankle activated and strong, the Achilles taut and ready to pass big forces from the ball of the foot into flexing knees and hips above. In slow-motion video, Marcus says, "It's like poetry. It's a stable platform; the force is being dissipated across all these joints in a balanced way."

And yet, as plyometrics became popular, most coaches—in the US, at least—coached to land toes first. "It was apparent to me that when athletes would land toe down, they would have a hard time getting into their hips," says Marcus. Instead, they braked in ways that stressed their knees. And through the process of landing and taking off again, they

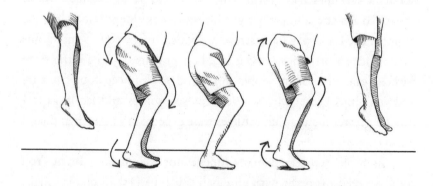

P3-APPROVED LANDING

seemed rickety. "It just puts you in a bad position to create force, and to move," says Marcus. "Super-unstable." Marcus compares landing toes first to "how a blind man might try to reach for the ground."

"And so, we were the contrarians," Marcus says. "Everyone was teaching toes-down landings, and I was telling all of our athletes and anybody that would listen that it's better if you land toes up, in a dorsiflexed position."

The biomechanics department at Stanford had taught Eric little about foot dorsiflexion during his major, and he didn't know what to make of his new employer's convictions about "toes up." But as P3's dataset matured, and Eric got a master's in data science and used new machine-learning techniques to help wade through the rows and columns, he found that the numbers backed Marcus's intuition.

The first finding was confirmation that athletes who landed with their toes up could get off the ground faster, and jump higher, than those who landed toes down. There are two big spikes of impact force in a toes-down landing: an initial impact spike when the toes hit, and then another when the heels slap down. "The amount of force there is inversely related to an athlete's jump," says Eric. "Generally, the bigger the impact, the lower the jump." The biggest impacts came from landing toes down.

A few years later, the dataset began to deliver insights that might prevent catastrophic injuries. Based on data about how NBA players moved *before* major injuries, he was able to quantify and rank many dozens of factors associated with some of the worst injuries in sports, especially ACL tears.

The presentation Eric made is thick with biomechanical and machine learning jargon and shows that "Max Hip Active Deceleration" (associated with 89 percent of injuries), "Femoral Rotation at Maximum Relative Rotation" (87 percent), "Drop Max Ankle Dorsiflexion" (68 percent), and "Relative Rotation at t0" (65 percent) played big roles. Also important was "Body Weight" (85 percent).

But one factor rose above all the others. Every single injury, a full 100 percent, involved something Eric named "Translation." "Nobody knows about that," says Marcus. It's a movement that hasn't been well

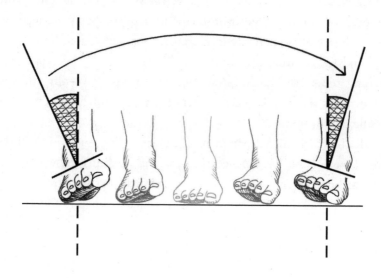

TRANSLATION

studied, perhaps because it involves a body part seldom associated with knee injuries: the foot.

"The video," says Eric, "is wild-looking."

Eric plays slow-motion animation of an NBA player's lower leg during a drop jump, with sensors stuck to four places around the foot and ankle. The foot arrives at an odd angle, and touches down—at first, anyway—only along the outside edge. It appears innocent enough. The trouble is that by the time the foot's edge runs out of elasticity, there's much more landing yet to come. As the video rolls, so does the foot, from the outside edge, across the sole, to the inside edge, "translating" the weight. By the end, the arch is crushed, and the shin bone has swiped, like a windshield wiper, from one worrisome extreme to the other. You might need a biomechanics degree to parse Eric's definition of translation ("axial rotation from the lateral aspect of the foot toward the medial aspect of the foot"), but you don't need one to guess that, off camera, the knee is unhappy.

Some ankle roll is normal—about eighteen degrees' worth of windshield wiper–style translation is typical in NBA players. As Marcus points out, if you land hard on the outside of your foot—called *inversion*—it's very common to then roll onto *eversion*. The force has to go somewhere. But in P3's data, when translation reaches twenty-five degrees, the red light of injury risk blinks on.

Translation is a chief obsession of P3. When NBA players visit Santa Barbara, they usually augment their training with daily sessions on nearby basketball courts, often with a skills trainer. (Not always, though. Marcus says Nikola Jokić declined the court time: "No, my skills are good," Marcus remembers Jokić saying. "Body bad.")

The P3 staff send players off to the basketball court with all kinds of known issues, from knee pain to tight hips. "I don't get concerned about that," says Jon Flake. But translation is different. "We can't ignore that," Jon says. When he's training new coaches, Jon tells them that if they see an athlete's foot striking the floor like that, even in warm-ups, "stop the workout."

How does translation threaten knees? Marcus tore his ACL when his foot was pinned to the ground in a bad position by a fallen teammate. Similarly, translation holds your foot to the floor, with a vulnerable knee, for a longer time than usual. Eric says, "Your foot rolling through big translation does not set the bony architecture in position to respond kindly to an ACL-threatening experience."

"Ground-foot interactions are super-important," says Marcus. "We tend not to give it a lot of attention."

Eric's finding about translation was a major revelation, but not the only one. Almost as shocking was the overall realization that knee injuries are seldom about the knee. The top *fifteen* factors in Eric's knee study all have to do with the hip or ankle. (The top-ranking knee measurement is knee flexion, with 17 percent impact, way down the list.)

"Where the knee is situated," says Eric, "it tends to be the unfortunate recipient of whatever occurs at the hip or the ankle. It's just unfortunate. You see some malady at the foot or ankle, more often than not, it will manifest at the knee." A dropped wineglass snaps at the stem because the heavy top half drives an unstable base into the solid tabletop.

Two hundred thousand American ACLs are surgically repaired annually, at a cost approaching $10 billion. Entire medical centers are dedicated to this injury. "Probably," Eric says, it's biomechanics' "most studied injury." Yet the standard source of insight is a scan of the knee, which, in the vast majority of cases, misses the cause. At P3, the struggle to prevent ACL injuries begins at the foot.

All in all, Marcus found the SIMI data offered "one big validation" of what his eyes had told him about landing. The data also shone light on the path to improving ground contacts.

In the leadup to the 2022 draft, I watched Eric assess a blue-chip Duke recruit on the table as half a dozen major NBA stars worked out. "We often say we are holding human potential in our hands," Marcus says with a grin. "Right now, our hands are very full!"

At one point, a crowd gathered. A mix of players and trainers *ooh*ed and *ahh*ed over something on a phone. "Look!" says P3 trainer Jack Anderson. "Look at his toes!"

The fuss was over a nearly decade-old video of soccer player Sanford Spivey jumping over a row of six red-and-black Retrospec-brand plyo boxes, one at a time. Typically, people jump onto or from boxes like these, but Spivey's not even touching them, flying clear over, landing on two feet on the floor, and immediately leaping the next one. They range from twelve to thirty inches tall. Spivey pops from the floor like a flea.

Shot from the side in slow motion, the video is frisky magic, hypnotic enough to be projected on the wall of a dance club. When Marcus talked to his daughter Kira's class, he says he "threw in that Sanford Spivey video," and it blew their minds, a class full of thirteen- and fourteen-year-olds reacting with "big *ahh*s," which is also what these NBA prospects do.

Spivey is what Marcus calls "a P3 baby, he trained with us since fourteen." Later, Spivey became a venture capitalist in Seattle. But at the time of the video, he was a Boston University soccer player, and Marcus says, "also pretty powerful. But mechanically, he learned how to jump perfectly." His feet are perfect, his toes are up.

"Typically," Marcus says, "nobody's ever taught basketball players how to land or jump." And they don't always want to hear it. Marcus has a story about Aaron Gordon listening carefully as P3 put him through a cautionary talk about how his left foot hit the ground. He came back a year later, having broken a bone in his left foot. "I was listening to you," Marcus remembers Aaron saying, "but I also said, 'Look, I hit my fucking head on the rim every time I jump, so what do I really have to change?'" The next summer, though, Aaron resolved to work on it, and Marcus says Aaron's ground contacts improved a ton.

Marcus sees this as a moment of change in agility and skill sports like soccer and basketball. He says sports like basketball and soccer "are becoming more and more ballistic." Since P3 began measuring, basketball has become so much faster that Marcus says "it's a different sport" than it was fifteen years ago. And this new sport is more like track and field, and happens at speeds where good landings are essential.

At P3, they show NBA players the Spivey video so they can see what's possible. They often also have them take off their shoes. When people do plyometrics barefoot, they naturally dorsiflex more. "It puts your body in a much better position to both receive force and then to deliver it. It also makes us a little quicker. Your ankle is loaded, it's ready," Marcus says. And that's the point. When an NBA foot hits the ground, who knows which way it'll go next? "Plyometrics train the system that allows you to just hit the button and go."

As force plates have trickled into athlete care, there has been an effort to minimize the big impacts of landing. But training to be an elite athlete is not about avoiding big forces. "If you want to go play in the NBA, or the English Premier League, you can't avoid dangerous environments. You need a system that can execute in this incredibly ballistic fashion. So, you have to put the body in the right position to do that successfully."

The ideal is not the avoidance of explosive violence, it's a shaped charge, pushing your body off in a helpful and manageable way. "Why is Kyrie [Irving] so great, right?" Jack says. "Kyrie is able to go one direction, put the brakes on, and immediately shift into another direction. *Perfect.*"

Training dorsiflexion is both a physical and mental process. (The Spivey video, as titled by P3, is called "Nervous System Training.") Jon says they change bad habits with aggressive strength work in the lower leg and a touch of brainwashing.

P3 programming is inundated with toes-up moments. Marching with a plate overhead? Toes up. Squeezing a medicine ball between your knees in a glute bridge? Toes up. First move of the standard warm-up? Walk a lap on your heels—toes up. "The more repetition of that," Jon explains, "the more likely we are to get some sort of adaptation to stick, so it's always a recurring theme."

Sometimes they quote legendary track coach Jim Deegan, who said, "Show 'em your sole," meaning lift your toes enough that people at the finish line see the bottom of your shoe. The coaching patter at P3 is thick with phrases like "come in loaded" to the floor, meaning to raise the toes and prepare the ankle to take the force of landing up into the Achilles.

Preparing for big forces like that takes time. Some steps are very basic: slowly raising onto tippy-toes and then back down, under control. Standing on one tippy-toed leg. Jumping rope with toes up. Sprinting. Professional basketball player Amber Melgoza first came to P3 after tearing her ACL in high school. Seven years later, after a workout preparing for her season in France, she sat on the pleather P3 couches. "You'll see a lot of people doing a lot of jumping," she says with a nod at the floor where NBA players were warming up. "And just the way they teach you how to break it down, baby steps and baby steps and gradually you can eventually start jumping on these boxes."

Every gym has different signals that an athlete is ready for aggressive plyometrics. At P3, they look for a dorsiflexed foot and minimal translation, after which they'll let you progress from jumping rope and dowel hops to box jumps. If you can do those with your hips and core in control, then you're on a path to the big line jumps that Sanford Spivey made famous.

Many NBA players struggle. Often, they spend more time watching another Spivey video, where he set a record that has held for a decade, hopping back and forth across the dowel forty-three times in eight seconds. That's more than five hops *per second*.

NBA players tend to start with far lower numbers, but improve quickly with practice, and the trainers feel good that as it does, the athlete's risk of knee injury goes down. Jon loves it. "You see big, athletic guys get exposed with just little hops. . . . They're so used to relying on

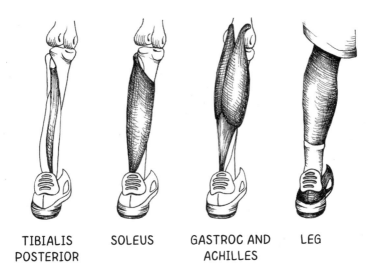

| TIBIALIS POSTERIOR | SOLEUS | GASTROC AND ACHILLES | LEG |

their longer levers . . . just focusing on the foot and lower leg is really difficult for them."

Another core component of P3's lower-leg regime involves three unheralded muscles, the tibialis posterior, soleus, and gastrocnemius. These three won't make the cover of *Muscle & Fitness* magazine, but they're critical to successful ground contacts.

The tibialis posterior (or, around P3, "post tib") runs deep through the lower leg and does the hard work of controlling the arch of your foot as you land. In other words, a strong post tib can dial back the inward collapse that is the scary second half of translation. A strong tibialis posterior can also dissuade the knees from falling into valgus position, knocking together like Kyle Korver's.

The gastroc is the bulbous thing you would recognize as a calf muscle. The secret soleus lurks beneath the gastroc, sometimes peeking out the sides. It generally holds your lower legs upright, preventing your shins from collapsing to the floor with every step. It's vital to pushing off in walking, running, or jumping. The gastroc and soleus weave together into the Achilles tendon.

The soleus also does something that I've heard doctors suggest might matter to tall people like NBA players more than most: it pumps blood back up toward your heart. A common thought is that the feet of the very tall suffer because their distance from the core inhibits blood flow. Many of the tallest NBA players—Yao Ming comes to mind—had careers end prematurely because of foot injuries.

P3 targets these three crucial muscles in myriad sneaky ways: standing on one leg while doing something with the resistance of a cable or weight, or raising onto the ball of the foot into a "march" position. One of the most athletic centers in the NBA, Myles Turner, "had been shut down with lower leg issues for a while," Jon says. Jon assigned him three rounds of five minutes jumping rope. Jon watched for short contacts on the ball of the foot. Five minutes was a dare—way more than the standard assignment of thirty seconds. Jon says Myles never completed the whole workout, but he got close, and his lower legs got much stronger.

In a video of Katie Spieler's ACL recovery, she's in socks, lunging with the front half of her foot on a low platform, and neither heel on the ground. Jon recommends three thirty-second sets of "heel hover split squat isos" to increase lower-leg stability. Another favorite: "foot bridge kettlebell pass." For this, Katie stands on just one foot, a heel on one ten-pound plate and forefoot on another. While doing all that, she passes a light kettlebell from her right hand to her left, swinging it out slightly to each side. She makes three sets of ten passes look easy.

At P3, you see quick, explosive things of all kinds: Hopping a row of hurdles a few inches tall on two feet, off a small box and then onto a bigger one, or leaping laterally across the floor onto one foot and then up onto a box on two. Pogoing on both legs up and down the track sideways, forward, and backward. Popping sideways off two feet clear across a lane of the track, and then back again. Starting with one foot on the box, and another on the ground, and then leaping up on top. Sitting on a box, and then standing and exploding onto another one. Stepping off a platform sideways, and then springing off the floor right back to where you started.

Jon says that a lot of P3's lower-leg exercises tax elite athletes, and "might blow out the calves." The next day they'll be sore and maybe walking funny. And maybe also the day after that. But Jon has learned that elite athletes' bodies soon adjust. By the sixth time that particular drill is in their rotation, he says they can handle it. And it's worth it. "You need good work capacity in the lower leg," says Jon, "to be safe at the knee."

"We do some hops, and then some bigger hops, and some bigger drops," Jon says of the progression. The staff scrutinizes the video, looking for toes to land up, weight in the midfoot, ankles under knees and knees under hips. "And we start to build up to these really big impacts."

"We find that if we get it right," Marcus explains, "we can really push on these systems." Jon agrees: "Aggressive jumps can make you so much more athletic and creates so much adaptation. It's useful to a basketball player."

Alex Ash, who runs the Lab, a P3 offshoot that serves nonprofessional athletes, said something that blew my mind: he has seen people with peak landing forces of twice their body weight—and he has seen people with peak landing forces *eight times* their body weight. Eric says he has seen *nine* times body weight. With forces this large, a little technique can make a huge difference.

The trick is to get those forces passed succinctly to the Achilles tendon. We say the best jumpers have "springs." It's literally true: the Achilles is an incredible spring, which feeds up the chain through quads to hips, all of which have important soft tissue. A lot of the exercise P3 prescribes is about positioning the foot and ankle so it's the Achilles that feels the force of the floor.

The mission is also to make the soft tissue of the lower leg *stiffer*. Not stiff in the sense of not being able to touch your toes (which a biomechanist would call "limited range of motion"). In biomechanical terms, *stiffness* is a measure of how much load a tendon or other tissue can take without trouble. "A stiff system," says Steve Magness in *The Science of Running*, "can utilize elastic energy better." (A rubber bouncy ball is a stiff system. A "dead"-feeling squash ball is not.) Studies show that runners who underuse their Achilles have to do 40 percent more work.

"Tendon stiffness would be the physical quality you're most directly trying to affect," explains Jon. "This is where the post tib and soleus come into play." Those muscles' critical job is to position the foot and ankle so the Achilles loads efficiently on landing. "If you contact the ground and collapse your arch, usually characterized by the navicular bone dropping down, then it's very difficult to have stiff ground contact." Simultaneously, a strong soleus stabilizes the heel. "Both help set up the foot," says Jon, "in a way that allows the Achilles to be stiff."

Even when athletes are training their larger muscles, P3 staff focus on feet. In coaching a squat, Jon tells a group of NBA hopefuls to find three points of contact: the big toe, the small toe, and the heel. He says to root the feet into the ground with a little outward twisting pressure. "If the foot loses composure, the knee goes with it," Jon says. "You want them to grab the ground, twist the plates, so they feel the tension in the hip and the foot." If you're doing it right, as you twist you'll feel your glutes come alive. "Especially in squatting," says Jon, "curing what the foot is doing will generally fix everything else."

# 12

## TEN MILLIMETERS
## FROM NOWHERE

*Begin at the beginning and go on till
you come to the end; then stop.*

—Lewis Carroll

ON ERIC'S COMPUTER SCREEN, James Harden's feet stabbed the ground. Eric rolled the video back and forth, as the accompanying rows of data delivered a mixed bag of insight. Most days, Eric's corner of P3 is as dry as toast, but on this day in 2016, things were a touch more glamorous. Not only did Harden himself sit two yards away, but also milling around were a handful of people who arrived with him, maybe a dozen more from Adidas and Adidas's ad agency. Harden was the best scorer in the NBA; everyone seemed certain P3's assessment would unveil exciting news of his athletic perfection.

The problem was, Eric had promised a little in-person preview of Harden's full results, and on his screen the data did not match the mood. "He's not," Eric says looking back, "winning a dunk contest." Compared to the NBA players in P3's database, Harden was "pretty average for his position," Eric says, "or slightly below in some respects."

Huh. Eric's mind danced. Harden had eluded the finest athletes on the planet, night in and night out, for years. Clearly, he could move. What were they missing?

Eric thought about how Harden's game was heavy with things P3 had never tried to measure: hand-eye coordination, shifty movements, reading the floor, and shooting. He's a master of getting opponents and referees off balance. Would that be enough to overcome mediocre athleticism?

Then Eric's brain made a conceptual leap—from a Houston Rocket to rockets in general. When a SpaceX rocket takes off, a crowd assembles outside the glass wall at mission control. They count down the launch sequence and burst into applause when the rocket ascends with a cloud of steam and dust.

But they're really waiting for the booster to return. "We used to just dump them in the ocean," says Eric. Now, as the SpaceX booster descends, the robot-driven engines fire *up*. "There's a big vertical acceleration," says Eric, "just shy of gravity." Instead of crashing into the launch pad, the booster hovers to rest like the emperor's ship arriving on the Death Star. In the final seconds, mechanical arms fold down and it parks, looking majestic. *That's* when the crowd goes crazy.

Bodies don't have jet engines that fire while landing. But they do have the eccentric movement of muscles. We think of muscles contracting—shortening, sometimes called "flexing"—which is what they do when we jump. Eccentric movement, on the other hand, is lengthening muscles under control. The force produced in eccentric movements slows you down in a controlled way, effectively pushing up, just shy of gravity, as the body comes down. "It's a physical quality," Eric remembers thinking, even if it's not a physical quality historically prized by All-Stars.

P3's general manager, Adam Hewitt, had prepared the team for Harden's visit with a dossier that showed, among other things, that Harden led the league in stepbacks. That's where Harden drives forward, gets the defense to worry he might keep going, and then stabs a foot into the floor, pops backward, and rises to shoot. It's a shot that works because Harden can change from forward to backward faster than the defense.

As a thing to unfurl to the assembled group, it wasn't too exciting, perhaps—but it was the truth. The secret of being James Harden, Eric nervously told James Harden, was something almost no one in basketball cared about: stopping on a dime.

Amazingly, the Adidas people loved it. "They really ran with it," remembers Eric. "Their excitement helped encourage us."

Eric says P3 then "actively started to look for, and studied, and tried to formalize" the biomechanical definition of what they call "decel," for deceleration.

Once they had defined the movement, P3 found elite stoppers everywhere: basketball, volleyball, football, soccer, and NASCAR pit crews. The data suggested that players with good brakes had meaningful advantages across sports. The data had coughed up a movement super-skill.

Decel is not assessed at the combine, it's not part of scouts' standard vocabulary, it's seldom mentioned on TV broadcasts. But it matters. With brakes as his only elite physical quality in P3's assessment, Harden won the NBA's Most Valuable Player Award the following season.

Just as Eric started to really dig into the concept of brakes, a seventeen-year-old Slovenian basketball player came to P3 for assessment while his mom sat on the couch. His data turned out to be a lot like Harden's: didn't jump, run, or cut especially well. But dang, could he stop.

*That's interesting*, thought Eric. *Let's see how this guy's career goes.* Luka Dončić quickly became the youngest MVP in the history of European basketball, an NBA lottery pick, and a perennial NBA MVP candidate. Other P3 All-Stars, like Zach LaVine and Darius Garland, tested with great brakes as well.

Suddenly, decel was a thing athletes wanted to learn, which gave Marcus a reason to get the whole P3 staff reading Verkhoshansky. There are two big advantages to training brakes with plyometrics. The first is that it grooves the feeling of a strong landing, upping the odds that it will be deployed in a game. The second is that it gets you moving into the next jump. Even in gymnastics, after flying off the uneven bars, you land, bend knees, then stand. In football, when you throw out a leg to brake, you then want to escape the cornerback with your next move. In basketball, when you step back, you want to jump into a shot. Jack explains that Verkhoshansky researched this exact thing, and "he found that if he just did the braking portion, but didn't couple it with some sort of concentric output afterwards, it actually robbed them of the natural movements you want to see."

Once their lower legs, and landing form, are strong, P3 trainer Jack says, he has players jump from a high box and explode into other movements. "There is no bigger eccentric stimulus we could give them," says Jack. By design, it's a recipe for big forces and a short amount of time. If the form is sloppy, that's precisely how injuries occur. With good biomechanics, however, those same ingredients inspire elite athleticism.

P3 has other ways to pump the brakes. Just as muscular tissue can be trained to shorten faster and more forcefully to lift an athlete off the ground, so can muscles be trained to elongate with strength against great forces.

For experienced weight lifters, the move involves an absurd, bar-bending amount of weight across the shoulders—quite literally, more than you can lift—and lowering it under control. The various techniques to end this kind of movement, one of which is simply ditching the bar, all require serious technique, which is why P3 never does overloaded eccentric squats with NBA players. "None of these guys," Jack says, "are great lifters."

A P3 workaround: Athletes hold a dumbbell in each hand, facing a box. Then, with deft timing, they dip down for a jump, ditch the dumbbells to the floor, and leap onto the box. It's a jumping exercise with weight added only in the eccentric approach phase. That adds training work, and puts a helpful stretch into the jumping muscles. Amazingly, the most common result is that players jump higher than with no weights at all.

Another favorite is a gizmo called the kBox, which looks like a high-end air fryer. In the most basic movement, you stand on the flat top of the kBox with a cable running to a harness around your torso. Then you squat. The kBox is different than pumping iron, however, because the resistance comes from a flywheel that sticks out the front of the machine. It's like winding and unwinding a massive yo-yo. The cable builds huge down-pulling forces you need to resist, but then reaches the end of its range, stops spinning, and then unwinds in the other direction, freeing you to stand back up. But you want to be thoughtful about that, too, because you're winding the thing for the next rep. "The faster I pull up," Jack says, "the harder it pulls down."

P3 coaches have found they can achieve about a 6 percent improvement in NBA player braking over six weeks. Players loved their new stopping speed. Adidas loved the new marketing message. The Harden Vol. 1 shoe launched in 2016 with a press tour that touted a "data-driven" pattern in the rubber of the sole, and the shoe included a rubber wrapper around the toe to keep your foot from flying forward when braking hard. It was positioned as a stopping machine.

But can a shoe really help you move better?

One April day in 2013, on an out-and-back of the Rutgers Unite half-marathon, I watched the lead pack scream past in the other direction. A man named Demesse Tefera seized the lead in a freewheeling way that moved me to scream, "HELL YEAH!" across the road. I pumped a fist, felt a rush, and sliced about three minutes off my personal best.

I can still hear the allegro cadence of his feet, alighting briefly and beautifully on planet earth, defying physics' technical term for the event: a *collision*.

In physics, there are two kinds of collisions. *Elastic* collisions involve no loss of kinetic energy. Picture billiard balls—one hits another, which departs at almost the speed of the first. *Inelastic* collisions get you a failing grade on the egg drop, because kinetic energy is lost and the objects slow down abruptly. Inelastic collisions tend to be hot and loud—the textbook example is a head-on truck accident. When Katie Spieler landed on her heels, it sounded like a sack of bricks and the force plates registered the impact of an angry bighorn sheep. (Related: "I heard a *pop*," people say, of tearing an ACL.)

The April footfalls of Demesse Tefera were barely detectable. He deployed some of humanity's best landing gear. Strong quads absorb huge forces, as do—in ideal running form—the soleus, glutes, gastroc, and hamstrings. Brilliantly, this system also stores those forces and can use them to spring forward again. Tefera's legs acted like rubber bands, pushing him forward with a blend of his own active work, and the residual freebies carried over from the last step. Tefera's footfalls were about as close to elastic as a foot-ground collision can be.

Tim Noakes's classic book *The Lore of Running* says up to 93 percent of the force of landing can be redeployed into the next step. Runners in cushioned shoes, however, pass along about half that. There's a chicken-and-egg question about whether we need soft shoes because we are bad at managing big forces, or we are bad at managing big forces because we wear cushioned shoes.

Harvard research led by Daniel E. Lieberman, Ph.D., finds that "habitually barefoot or minimally shod humans tend to walk and run differently than shod people, often in a way that leads to very low collision forces, even on very hard surfaces." After a lifetime in shoes, Lieberman began running barefoot around paved Cambridge, pointing out that, evolutionarily, it's wearing shoes that's weird.

What seems to be happening is that the relatively small amount of padding in a shoe sole makes it feel safe to land on toes, heels, or any old place. But humans are heavy eggs and that rubber is thin, with a minuscule fraction of the shock-absorbing powers of Demesse Tefera's collection of springy muscles. Without the body in a good position to deliver the landing forces to the quads and glutes, the landing becomes inelastic, like the truck crash in the physics textbook. And remember how crashes make noise? They also emit heat. Noakes writes that "the remainder of the energy heats up the midsole."

Think about it: you're killing yourself to move forward, and a good chunk of the force you're creating is being lost to the silly side project of simmering your expensive shoes. Imagine a Formula One racer losing power to a toaster oven in the cockpit.

There's precious little academic research on basketball shoes and safety, but a 2017 article in *Lower Extremity Review* quotes an academic as saying shoes "can definitely make things worse." Describing a 2022 survey of running shoe studies by *Cochrane Reviews*, the *New York Times*'s Cindy Kuzma wrote, "the analysis found no evidence that running shoes, or prescribing certain shoes by type, have injury-preventing properties."

Shoe companies frequently commission P3 to consult. But to

P3, biomechanics often feel like an afterthought in those discussions. "They'll draw up something they think looks really cool, in quite a bit of detail. And they'll say that's gonna be Ant Edwards's shoe and then they will have their tooling guys make it. And it's really starting from the outside in," Marcus says. "It's done with the premise that they have to build something that's cool to sell to millions of kids. And I get that.

"But I just keep saying you've got to start with functionality. You got to think of Ant Edwards as this amazing machine, and you're going to build some tool for that machine that's going to make the machine work better." On the day we spoke in 2024, Marcus had just visited Adidas in Portland, Oregon, where he says he pitched an entirely new way to design shoes, which would mean having the best players in the world move around on P3's force plates in several different prototypes, while Eric's team assessed to see which, if any, offered movement advantages.

If that process produced an athletic shoe with a real, demonstrable performance advantage, it might be the first of all time. Marcus and P3 have a front-row seat on how shoe company research money is spent; some of it is spent at P3. Marcus is surprised at how seldom P3 is asked to look into the big questions of sports, like how injuries correlate with shoe design, what body type is best in which shoe, or whether shoes change how athletes move. Marcus says, "They're mostly about colorways and design features."

In the *Lower Extremity Review* article, representatives of the major shoe companies tell writer Will Carroll that when shoe companies call a shoe "custom" to a player, that doesn't mean it was designed for his foot. It means the player told the shoe company what color and level of cushioning he likes. What tends to happen is that players like something that the shoe-buying public also likes: that squishy-cushion feeling when you try it on. ("It's like sex on your feet," Joakim Noah once told me about the Boost material in his Adidas, at an Adidas promotion event, sharing a couch with the Adidas CEO.)

Noakes calls athletic shoes an "expensive gimmick" in need of an infusion of hard-hitting research. "Perhaps," writes Noakes, "just as the pharmaceutical companies must expose their drugs to extensive testing before they may be released for use by the general public, the time has now dawned for running shoes to be subjected to similar considerations."

Meanwhile, you know who wins a laboratory treadmill test, running as fast as Usain Bolt's world-record hundred meters thirty times in a row, up an 11 percent gradient? A pronghorn antelope, that's who. And they run on goddamned hooves. (This actually happened. Scientists put pronghorns on treadmills.)

The best athletes P3 has ever tested were not antelopes, but they did have something in common with them: instead of needing protection from the force of landing, they used the landing collision to begin the next bounce. Demesse Tefera and Usain Bolt know that human bodies can carry energy from one movement into another without cooking our soles or making loud slapping sounds.

Unfortunately, in that project, shoes often get in the way. Equipment-wise, basketball doesn't take much. All anyone really needs to ball, besides a ball and a hoop, is sneakers. The cost of shoes is nothing compared to a bag of hockey equipment or a set of golf clubs. But it's enough that the athletic sneaker industry generates about ten times the revenue of all the teams of the NBA combined.

Yet, every day, P3 staff asks athletes to take off their shoes for certain parts of the workout because wearing shoes might *increase* the likelihood of catastrophic injury. NBA All-Star Khris Middleton has worn Nikes since college, but at P3 he does trap bar dead lifts with no shoes at all. Major League Soccer's Juanjo Purata does aggressive hip flexor work against the

10 MM

HEEL DROP

resistance of the cable machine in black socks. When the WNBA's Haley Jones slipped a green exercise band around her foot before a predraft hip exercise, her shoes were on the floor by the couch, next to her phone.

Marcus often rants about footwear. His first concern: athletic shoes tend to have a ten-millimeter drop from heel to toe. "No one," says Marcus, annoyed, "can tell you why it's ten millimeters. There's no magic about it. Lifting somebody's heel, you affect lots of mechanics—more in crappy ways than in good ways." Certain kinds of weight-lifting shoes have elevated heels to allow for deeper squats, but other than that, P3 sees no benefit to elevated heels.

In fact, they see danger. Elevated heels encourage people to land toes down. At slow speeds, that increases forces running through the knees. In big, explosive movements, that "puts their knees in a dangerous position," explains Marcus.

Marcus notes that some of the edgier running companies have sold zero-drop shoes since *Born to Run*. "I've been advocating this," Marcus says, "for a long time." Though Marcus, like many athletes, has an extensive collection of standard ten-millimeter-drop shoes, he often wears zero-drop shoes, or Birkenstocks. He says he's happiest barefoot. (After a summer trip to Venice for the Biennale, a citywide art exhibition, Marcus reported loving the people, the setting, the food, and the fact that "I did everything barefoot.")

Marcus may not be able to tell you why basketball shoes have a ten-millimeter drop from the heel to the toe, but Dave "Boot" Bond can. Bond was the director of basketball at Nike when they created the Jordan, Pippen, and Barkley signature shoes. He wrote the original business plan for the Jordan Brand and hired the team that launched it. He has played similar roles at Adidas, Anta, Quiksilver, and UnderArmour.

"I've spent my career in the athletic footwear industry," Bond says. "I got to work with pretty much the entire Dream Team, including Michael Jordan. We made his shoes. That was high stakes! Making footwear for the world's most famous athletic person. You couldn't goof around and make things that just look good."

In the explosive early days of athletic shoes, though, there was a gimmick that reliably moved product: a pocket of air in the heel. "Back in the day," Bond says, artfully avoiding naming a company, "you had to fit a visible air bag and had to make room to fit a technology. It became the standard look and silhouette of a basketball shoe."

*That*, Bond says, is why athletic shoes have a ten-millimeter drop.

A fixed height in the back, Marcus says, means the same model of shoe has a low slope for Shaquille O'Neal, who wears size 22, and a steep slope for a small child. "Those are totally different angles, different dynamics, different physics," says Marcus. "How can they both be right?"

Bond says that there's "no scientific basis for it."

In 2006, Bond joined K-Swiss in Los Angeles, where his first big project centered around eye-poppingly athletic French tennis player Gaël Monfils. Monfils made highlights leaping, cutting, and murdering the ball in exciting ways. "Somebody told me," Bond remembers, "about some guy in Santa Barbara doing cutting-edge biomechanics with athletes. They told me all about him, and I am stupid that way—I just drove up there."

Bond walked into P3 with no appointment. Before long, Monfils was in Santa Barbara, too. In 2011, K-Swiss had a hit with the Big Shot shoe featuring something called a "Ballistic Propulsion Plate" under the ball of the foot. Marcus still has a Big Shot in his office, complete with a P3 logo on the outside.

What Bond remembers most is the first time Marcus really unfurled his cavalcade of sneaker industry complaints. "He was never mean," Bond says, "just frustrated, having to take athlete's new shoes off, to put more generic shoes on them to do testing. The products were going in the wrong direction, he said.

"Of course, he blamed me for the entire industry. So, I said first of all, 'I didn't do that.' And second of all, 'What would *you* do, smartypants?'

"He said, 'Your industry doesn't realize that force is your friend.' The force that an athlete generates landing—great athletes use that force."

Bond was sold. In June 2016 *Sports Illustrated* featured Marcus and Bond's new shoe brand, Ampla. The running shoe featured a carbon-fiber plate in the sole. "The model when building running shoes has been that force is your enemy, this the opposite model," Marcus told *SI*'s Tom Taylor, who wrote that "Elliott wants to redirect the force, not squish it into a wedge of foam."

It was halfway to a pronghorn hoof, and inspired by the blades worn by amputee sprinters.

Marcus wanted it even springier. "I wanted to start with a pogo stick," he says, looking back, "and then just tune it down a little." At the end of the *SI* story, Bond says something that sounded kooky in 2016: he hoped this kind of carbon-plate running shoe technology would lead to someone breaking two hours in the marathon.

Marcus still wears an Ampla hat to work sometimes, and an Ampla shoe is one of the three or four "decorations" in his bare-bones office. But the shoe hit the market with terrible luck: Ampla's parent company, the surf brand Quiksilver, declared bankruptcy in September 2015. Bond, Marcus, and a couple of other investors bought the rights to Ampla. Bond rented an office in Los Angeles, where he personally processed and shipped the orders as they came through the website.

"We sold about a thousand pairs," says Bond. "Nike's head of innovation bought, like, twenty pairs. A few years later, they come out with the sub-two marathon shoes, which are ridiculously similar. We started a conversation in our industry about hyper-performance. They took the germ of an idea and really spread it. I would say Marcus is one of the inspirations for carbon-plated shoes. He wanted shoes to be not mushy and soft, but spring-loaded."

On the feet of Kenyan distance runner Eliud Kipchoge, Nike's carbon-plate shoes lit the record book on fire. At the 2018 Berlin Marathon, Kipchoge broke the world record by a minute and eighteen seconds, with a time of 2:01:39. In 2019 in Vienna, assisted by a team and a pace car that invalidated it as a world record, Kipchoge broke two hours, running a once-unthinkable 1:59:40. In 2022 he took another thirty seconds off the world record.

Running shoes with carbon-fiber plates have created such a tectonic shift that legendary runners like Kara Goucher and Des Linden spent an hour of their podcast explaining why records from before carbon plates are no longer relevant. Times have fallen in every event. Goucher likens shoes with carbon-fiber plates to doping.

One measure: in the fifty-seven years after Jim Ryun became the first high-schooler ever to break four minutes for the mile, another high-schooler ran that fast, on average, every 4.75 years. In 2022, when "super shoes" hit the scene, six different high-schoolers shattered the barrier.

The running shoe market, with carbon fiber and an array of zero-drop options, is showing signs of evolution. Basketball shoes, on the other hand . . .

"There are a handful of flat-to-the-ground running products," Bond says, "but the whole barefoot movement in running never made it over into basketball."

"They're not as comfortable," says Bond, "and they look weird." Basketball shoe purchasers, Bond says, "want performance right up to the point it doesn't look weird. These shoes are bought by kids, literally twelve- to twenty-year-olds. If you're beyond twenty and playing basketball, you're old."

Bond joined UnderArmour in 2018 and says, "The Curry 8 and 9 are heavily Marcus-influenced, even though he doesn't know it. I consider myself a disciple of P3 and I've applied those things. Those are probably the most P3-centric shoes on the planet. They're not completely flat—there's a six-millimeter heel lift, I believe. So, it's 40 percent closer to the ground than most." Bond says Curry had a career resurgence around the time he started wearing them, then adds, with a laugh, "I'm not going to take *all* the credit." That's not to say the shoe industry has truly applied the lessons of biomechanics generally. Eric works alongside Sloan Hanson, a biomechanist who grew up playing soccer and working out at P3. Sloan studied data at Berkeley, and now enjoys exploring how shoe models correlate with injury in the NBA. Some of his early work found that three models made under one of the biggest athlete-name brands led the league in injuries per game in 2023–2024.

"Because we have so much data on some of these athletes, we could actually start looking at some of those patterns, their movement patterns, associated with guys playing in Kobes versus LeBrons. They should know these things, right?" Marcus says. "They should know if [any particular shoe] decreases your risk of injury or increases your risk of injury."

In P3's assessments, Marcus feels like he sees players moving in unnatural ways that would have never developed if they had grown up barefoot, or in very different shoes. "When guys do slides in our facility," Marcus says, "they stomp on their shoe when they change directions. They treat it like it's a big berm." Baked into thinking about basketball shoes is the assumption that the foot lands on a squishy platform, and then, Marcus says, "you try to figure out how to keep the foot attached to the platform." But what if, instead, he asks, "you made shoes that were more wrapping the foot?" Would basketball players move like ballet dancers? Would basketball players, Marcus wonders, land, cut, and jump in ways where "your foot is still operating like a foot"?

Trent Reeves is a biomechanist, not a shoe designer, but he did work at one major shoe company for a year and he joined many of P3's meetings with other shoe companies. "My experience in footwear has been [that they design footwear] kind of backwards," Trent says. If a shoe company has some technology that's selling well, he says, using that becomes the mandate, rather than asking how a shoe might, say, prevent ankles from rolling inward.

The P3 ideal would be for shoe companies to start with some performance or injury-prevention goal. Eric has poked around in the data already on P3's servers. "Talk about low-lying fruit," Eric says. "If they wanted to get serious, there's a lot that could be done."

Should different people have different kinds of shoes? Marcus points out that there are special shoes for male guards and for male big men, but there isn't a single basketball shoe designed for a woman. "Women's shoes," he says, "are small men's shoes." The most-worn shoe in the WNBA is a Kyrie Irving model. How, I wonder, would Marcus make shoes differently for women? "That," Marcus says, "is what we need to find out!"

"At some point," Eric adds, "it will really benefit the world's greatest athletes to put 'em in shoes that really suit them." P3 already sorts athletes by categories that emerge from studying movement patterns, performance, and injury risk. It would surprise no one at P3 if the right answer proved to be shoes designed not for guards, big men, or women, but athletes who land with toes down, in translation, or with feet that tend to rotate on landing.

One shoe design for everyone feels, to Marcus, long outdated. Breast cancer researchers are decades into seeing that survival improves dramatically when you dig into the specifics of each case—HER2, hormone-sensitive, there are dozens of varieties that suggest very different therapies. Studying all cancer patients together, as a monolith, would obscure the most effective remedies. Marcus suggests the same will prove true for athletic shoes. "This old analytical model we have of just aggregating data and seeing what the net effect is, misses so much signal," he says. "That's true in drug discovery, surgical interventions, poverty studies. So, without question, there are some people that are going to be much more dangerous running in a minimalist shoe. They need some things that a cushioned, lateral stability–based shoe can provide. They basically need some help."

In 2017, big footwear companies announced they would scan individual runner's feet and then build shoes for them in "fast factories." But pretty soon, Marcus says, "they realized that they didn't know what questions to ask to build a shoe for you. Other than get a 3D scan of your foot." Once again, it was the problem of a static scan versus a study in motion. Robots might be able to solve the relatively simple problem of making a shoe, but they couldn't solve the wicked nuances of human movement.

"What about your system? Where do you need help? You know, how are they going to optimize you to reduce the risk of injury?" Marcus says. "I was super-excited about it, and then I realized that they . . . had the technology to build it, but they were not actually very close to having any knowledge to build it."

Meanwhile, basketball players don't have much choice. Trent is the star of P3's local rec-league championship-winning team. I ask him what he plays in. "I play in Dellys." (Did you know that longtime NBA role player Matthew Dellavedova has a signature shoe?) I asked why. "Because he's a really cool guy," says Trent. "Because he came in and we trained him for a while and he was nice enough to give us some comps. And so I got his Santa Clara version of the shoe. But yeah, it's funny: when I play basketball and put on a pair of basketball shoes, like the brain, the science brain kind of turns off a little bit and it's just, like, 'These feel good. I'm gonna play in these.'"

# 13

## DRAGON'S BACK

*Dragons and snakes aren't so different.*

—A. D. Aliwat

ON REPORTING TRIPS to Santa Barbara, I'd pack only a small back-pack, rent a bike instead of a car, and wear one pair of minimalist, zero-drop, thin-soled sneakers for everything. Marcus called me a Santa Barbarian, emphasis on the second word.

My super-light Inov-8 shoes were fine for biking around and inter-viewing people, good enough for whatever adventures might arise, and perfect, it turns out, for climbing Dragon's Back—the funkiest part of the ridgeline above the city, a well-named nobbled crest.

That weekend day began with a Marcus text about a hike. As I replied with a thumbs-up, I was not feeling so barbarian; I was flop-ping around on the rug, hoping to identify a stretch that might address an oddly stiff lumbar, with a sidecar of a little electric tweak down my right leg.

A few hours later, as Marcus pulled the Audi over near the trail-head, it was honest August midday desert heat.

Five of us wound along the edge of a canyon, and then darted up through trees. Before long it was the kind of steep where you can reach forward, without bending, and touch the rocky ground.

Beneath the tree line, the scramble felt like hot yoga. Higher up, it became rock climbing on toaster ovens. Much of the "trail" is made up of white arrows up sandstone boulders. The sun baked the rocks and the rocks pinkened our palms. Our sweaty shirts picked up stone dust and our eyes picked up views of gliding grackles, wispy clouds, and a glistening Pacific. But my thin shoes gripped like climbing shoes.

Many of the shrubs are bay, as in bay leaves or bay rum. Fancy hotels wish their lobbies smelled so fragrant. It's a vertical desert. Soil and water barely exist; instead, there is rock, rock, rock dust, and bay. And a few improbable, scrubby pines. "They're like NBA players," says Marcus. "It's a horrible environment, but here and there a few can adapt and hang on. The survivors."

As we hiked, Marcus brought up a work conundrum: James Harden. In the peak of his career, defenders essentially stopped trying to block Harden's shot, for fear he would win a referee's whistle with clever feints, leans, and pumps of elite brakes. Harden led the NBA in free throws made for six straight years. But by the summer of 2022, defenders were blocking his shot and barking about it. After so many years as a predator, he had nights as prey.

Adidas, which had invested hundreds of millions in the former MVP, brought Harden to P3 for his third visit on Tuesday, August 2. Eric had been excited about it. Bubbly good news flows easily at P3, I expected I'd hear all about Harden when I arrived the following day. But for most of the week, the name Harden didn't come up. On Friday, though, a high-schooler jumped twenty-three inches and somebody quipped it was "higher than James Harden."

In those years, Harden churned through squads—the Rockets, Nets, 76ers—without too much winning and with reports of a balky hamstring. The 2022 offseason, though, offered a chance to make things right. He took a pay cut, signing a nevertheless-lucrative extension with the 76ers, and was finally getting himself ready to be a key part of a contender.

But Harden hadn't stuck around to hear the results of his P3 assessment. Instead, he scooted back to LA, where he was mired in pickup basketball. Video leaked; especially two straight plays where Harden put the ball on the floor to attack twenty-year-old Toronto Raptor Scottie Barnes. Barnes is one of the best defenders in the world, young, and mega-fit. (When he won Rookie of the Year, Barnes sent P3 a hat that he inscribed with "You guys are the real MVP.")

In one clip, Harden puts on a ball-handling display, but can't get around, over, or through Barnes. When he gives up, shoveling the ball to an open teammate, Barnes claps with glee. In the second clip, Harden pivots, fakes, fails to get Barnes to fall for anything, and lofts an air ball.

None of this surprised Marcus. Harden's assessment data is proprietary and Marcus won't share. But it's clear that Marcus thinks that working with P3 would have helped Harden regain his prime form. Marcus says that if he did the right work, Harden could have two more seasons of moving like he was 26. Alas, neither Harden nor his team know what's in Harden's P3 file.

Shade came only in tiny nooks where a pine met a rock face. We tucked in there as if being hunted by drones. Now and again an ocean breeze would drift by. Marcus saved a few ice cubes in the bottom of his Hydroflask. After the water ran out, he gave us each one cube, a desert sacrament.

The problem, I suggested, is that Harden doesn't trust anyone who can deliver Marcus's insight. (Harden was about to have a falling out with Morey, and his agent, too. He'd soon join his fifth team, the LA Clippers.)

"Maybe," said Marcus. "I think the problem is that he just doesn't know." He just doesn't know what granular movement data can do for him.

Short of the final peak, we declared victory and turned around. On the way down, we found two young women cowering in a sliver of shade. One stood, concerned, over the other, who slumped. Marcus had just been telling me an insane story about carrying a large injured man down this mountain. Was something similar about to happen?

Marcus leaned in. For five minutes, they discussed the route, the heat, where they'd parked, and how much water they had left. Marcus said reassuring things about how much faster it would be going down, how close they were to the trees and the shade. A little later, he noted they were still sweating, talking lucidly, and had water—thus his conviction that they wouldn't require help. Thoughts of Harden melted away.

Before dropping into the welcoming shade of the woods, Marcus sat to drink in the view one last time. His LOVE YOUR NEIGHBOR baseball cap was off-kilter and sweat-streaked. He put an arm around my shoulder. "You don't see a lot of sixty-five-year-olds up here," he said. "But in ten years, we're going to be back here, and it's going to be glorious." He has a gentle way of throwing down gauntlets.

A half hour later, in the juniper quiet below, a burst of pings announced that the phones in Marcus's string backpack had reentered cell phone range. Marcus took a glance and said, "Oh wow."

It was the longest text I had ever seen. Star point guard Jrue Holiday was due at P3 the following day. His coach at the time was Mike Budenholzer. Before the hike, Marcus had invited Mike to visit P3 to see Jrue's visit. This long text was Mike's response. But it was not about Jrue, athletics, P3, or the Bucks. It was about Mike; his back hurt.

Budenholzer isn't the only NBA power broker who trusts P3 with their own knees, backs, and hips. The sons of Sacramento Kings head coach Mike Brown, legendary player and high-ranking league executive Joe Dumars, longtime general manager Mitch Kupchak, Indiana Pacers billionaire Herb Simons, and former Cavaliers coach David Blatt (whose son Tamir, Adam notes, is a good player) all trained at P3. Danny Ferry sent his son and daughter, as did Kings investor Mark Mastrov, founder of 24 Hour Fitness. Clara Tsai, who's married to Brooklyn Nets lead investor Joe Tsai, had her visit put off by the pandemic. Do you know who the strangely intense Memphis Grizzlies billionaire Robert Pera sent to P3? Himself.

We made it back to the car with tongues hot in our mouths. Any liquid would have been good, but we shot the moon, drove downtown, and walked to a smoothie place called Blenders. My lower back had been pain-free all day, but started to feel strangely tight, and a little asymmetrical, after climbing out of the car.

Sitting on the sand of the Jersey Shore a week later, my brother-in-law asked about the bony bulge on my lower spine. The truth was, I had no idea what it was. I had been asking around. A chiropractor had tried some things. I had an appointment with a sports medicine specialist.

I tried to keep moving. The next morning, I got up early and walked to a schoolyard to exercise in the pre-scorching hours. A kindergarten-aged girl and her unshaven dad shot hoops. Elbow tucked, release snappy, eyes locked on the rim, he drained a crisp free throw. She asked why his shots went in and hers didn't, and he said because he used to play basketball. He might have been forty.

His next shot pinged off the back rim and looped over the blacktop on a tangent every player knows: hustle and it's yours. Or amble, miss the cutoff, and the ball bounces to the great beyond. He yelped with his first step. His shoulders fell. He wheezed and planted his hands on his thighs. Back pain. His daughter took off after the ball, slowed, and then got distracted by some flower or bug.

People his age play in the NBA. *What are we doing to our bodies?* I wondered a little smugly. Then I began a workout carefully modified not to tweak my own little *situation*.

Over the weeks to come, I couldn't shake the twinge. I swam, I surfed. I drove to the good yoga place, lowered myself into ice baths, and slept on a heating pad. An orthopedist took an X-ray from the front and said it was hip arthritis. The sports medicine doctor took a better X-ray from the side, noted that the lump was because a vertebra had—really?—rotated into a new, incorrect position. He offered a prescription version of ibuprofen called Celebrex, as if there were something to celebrate.

Marcus is determined to prevent musculoskeletal injuries, to keep people from going through his ACL funk and its aftermath. Against the impending modern doom of immobility, which robs joy from life, his mission feels enormous. "This is all we have," Marcus once said, leaning an arm around Nadine and waving a hand at their bodies. "Without this, it's hard to be happy. Movement is everything. Movement and relationships."

Sometimes at P3, they talk about *the point of no return*, where a body gets too unathletic and pain-ridden to really get moving again. With World Health Organization stats calling immobility the globe's fourth-leading cause of death, it doesn't feel safe to wander near that point.

In the car, my high school–aged son Duncan was eager to tell me what he had learned going down a rabbit hole on epigenetic aging, which is a way to assess the wear and tear on a body's DNA. We're wired to move, he explained. And once we stop moving, we are wired to die. Harsh! But plausible in the research. Physical activity slows epigenetic aging; inactivity speeds it up. What nomadic tribe of hunters and gatherers could sustain a big population that neither hunts nor gathers?

In October, the ding-dong chime of nerve pain summoned Satan himself. That redneck's chainsaw ran up, down, and all around my lower back, hip, and right leg, shredding my old pain threshold. The data shows that being unable to stand without the assistance of your hands predicts an earlier death. I couldn't *sit up* without them. I canceled everything fun. "Pain is not just bad in itself," writes MIT philosophy professor Kieran Setiya. "It impedes one's access to anything good."

When I told my primary care physician my tale, she asked if I would like to do what every doctor would recommend: physical therapy. One damp morning, sunshine glinting off leafy puddles, I watched a driver pull into a parking spot at the physical therapist's office and kill the engine. Then—other than some highway *whirr*, the tick of a cooling engine, and the chirp of a bird—nothing. No door opening. No getting out.

And no mystery. I knew what was happening, because I was in park, too, summoning the strength to swing a leg to the ground, knowing that would crack the whip of nerve pain. Exhaustion from lack of sleep, creased faces, nonskid shoes—this parking lot is an assembly point for bodies of regret.

"I move," writes novelist Haruki Murakami, "therefore I am." But we couldn't move.

Nor talk. In addition to everything you'd imagine, the pain or the pills or something had dimmed my ability to use words. Writing became tricky. Phone calls became exercises in concentration, and small talk an absolute minefield. On TrueHoop's podcast, I'd arrive at mid-sentence wondering where I'd put the end.

"Physical pain does not simply resist language," writes Elaine Scarry in *The Body in Pain*, "but actively destroys it, bringing about an immediate reversion to a state anterior to language, the sounds and cries a human being makes before language is learned."

*Uh-huh.*

As winter blew in, I stopped running, surfing, CrossFit, yoga, and driving more than fifteen minutes away unless absolutely necessary. One November day, I made what felt like a bold plan: I would ride a bicycle. Not far, not up a hill, and nowhere I might fall. I pulled the mountain bike carefully off the bike rack by a beautiful flat path on an old railway line along a river.

As he fought lung cancer, my dad obsessed over rides like this. On the last one, I shot a little video, from my bike to his, as he thanked me for bringing him to the woods. "I just love it," he said, cheeks wet with tears.

For the months prior, healthy people stepping, pivoting, laughing, coughing, or bending seemed to be showing off. As in: *It feels so amazing to move!* Of course, they weren't thinking that, they were just doing regular stuff. But on this bike, every air molecule hit my face with stirring magic. The churning knees! The shifting clouds! *Of course* bears frolic after hibernating. The last six letters of *emotion* aren't random.

As I rode, I listened to academics on a podcast, explaining that we are evolved from fish. Touch a fish on its left side and it'll zap its entire self to the right, so lightning fast, it doesn't even use its brain, which would be too slow. Instead, motor neurons in the spine orchestrate the job. Fish ballistics.

Humans never lost those neurons, but by adding arms and legs we made movement lot more complex, brainy, and slow. Fish save their lives with their spines. People save lives—in movies, anyway—by driving a car, grabbing a gun, or defusing a bomb. It's all fingers, things at the end of limbs—as we grow weak, immobile, and uncoordinated in the middle. There's neurological work to moving a body as complex as this.

Our brains evolved to do that work. Academia confidently omits bodies when cataloging human genius. The word *brainiac* applies to people sitting still, playing chess or reading a book. But our brain's core work is moving our bodies. British neurologist Oliver Sacks says that "much more of the brain is devoted to movement than to language. Language is only a little thing sitting on top of this huge ocean of movement."

In November, a physical therapist named Jen talked me through the radiologist's MRI report: Herniation. Diffuse disc bulge. Moderate-to-severe bilateral foraminal stenosis. "Possible contact upon the traversing L5 nerve roots." Diffuse disc desiccation. And Grade 2 anterolisthesis, which is where something called a *vertebral body* slips out of position. Later, I learned that common symptoms from that last one include "sharp pain, numbness, difficulty walking, muscle spasms, postural issues, limited body movement, loss of bladder/bowel control, loss of sensations." By that point Jen, in a chair near my head as I lay on a heating pad, still wasn't done reading all the findings. She summed it up as: "You have a lot going on."

I home-brewed a new bedtime regime: Celebrex before bed, instead of with dinner, and it might work until sunrise. Don't hit the bedroom without the big, flexible, terribly cold ice pack, a heating pad, and the Theragun. Sleep on your left side on a foam pad on the floor, with two pillows beneath the head and a third between the knees. If you're crafty, you can dock in there between the pads and power cord just so and end up with blazing heat on the lumbar and brutal cold to break up the nerve-pain party on the right shin. And don't forget the Costco bottle of Extra-Strength Tylenol, emptied into a handmade ceramic bowl, near the pillow.

I learned to nudge my torso around, chest down, an inch at a time, through lunar phases, to switch off the lamp. It wasn't perfect, obviously. The dark silence of our bedroom was often shattered by the Theragun's manic whine. (John Lennon's song "Whatever Gets You Through the Night" and I were both born in 1974.)

Of course I missed romance, running, restaurants, tying my own shoes, sitting in chairs, and fun. But one thing at a time.

"Oh my God, Henry," my brother-in-law John said from the bedroom doorway. They were sleeping over for the holidays. He was seeing my new setup for the first time. "This is so sad!"

At once, I could see my life through John's eyes: me, immobile in the dim light next to the dog bed, the worst-round draft pick to write a book about the joy of movement.

In January, pain specialist Sameer Siddiqi, DO, pulled my MRI up on a computer screen and zoomed around like a TikTok of a roller coaster. Mostly, you could see healthy white nerves, like harp strings, dangling down the dark tunnel of my spine's interior. At a certain point, though, even a noob like me could spot the glowing knob of disc tissue poking in. "Touched a nerve" is a saying.

It was *a* problem, and perhaps *the* problem. A huge number of people have back issues that show up on MRIs but simply don't cause pain. The trick isn't to address every little thing, but instead to address the actual cause. We had a target.

Even as Dr. Siddiqi delivered my first epidural injection, he kept up a sunny banter about my path back to CrossFit and running. I'd make it, he thought. The injection would get me into more meaningful physical therapy, which would lead me back to the careful exercises in the gym that would . . . well, that's where it got tricky.

In the dark nights on the bedroom floor, I had a recurring fantasy of ocean swimming. In my warm, small stillness, I imagined the roll and whoosh of vast cold. Would it feel better to be a fish? I called a friend; on a thirty-three-degree morning, we watched the sun rise over the Atlantic while bobbing in wetsuits and fifty-two-degree water. Then we went to work and texted back and forth about how fantastic we felt. But it didn't last.

My MRI unlocked precious health-care dollars. Surgery and opioids were mine for the asking. The system works for patients lying still; the highest doctor in the land is the *surgeon* general. But only in the MRI tube or anesthetized on the table are humans so static. In the real world, we lean, wiggle, shuffle, hop, and bob in the sunrise.

Where could I find someone who could see what the scan missed: the hip movement that caused the entire ruckus?

# 14

## GET INTO YOUR HIPS

*It's a hell of a project.*

—Eric Leidersdorf

"HAVING HIPS THAT work is so essential," Marcus says, "to having a body that works well."

But we don't. Almost every athlete P3 has ever assessed has hip issues—and the rest of us are generally hip-illiterate.

We understand that the Federal Reserve can do something in New York that changes prices in the dairy aisle. It makes sense that acres of pavement upstream can cause flash floods three towns down. We are only just beginning to comprehend that problems in the ankle, knee, or back often derive from hips.

In 2012, Stephen Curry was not a huge star. His first three years in the NBA had been more Job than job; mostly, he spent time fussing over his chronically sprained ankles. A surgeon told *ESPN The Magazine* that the inside of Curry's ankles looked like "crab meat."

At twenty-four, Steph should have been ascending to his peak. When he finally returned, his teammates noticed that he avoided hard contact and vigorous cuts, while drawing fewer fouls, grabbing fewer rebounds, and blocking fewer shots. Retreat mode would never beat LeBron James, who looked, physically, like two Stephs. In 2012, LeBron won his first title, and then in 2013, another. The Golden State Warriors had a slender, beat-up, would-be star and the NBA's tenth-best record.

That's when the Warriors hired Keke Lyles, a strength coach with a doctorate in physical therapy and a deformity of his femur. Even before he got the job, Lyles had ideas about fixing Curry's ankle problems upstream.

Steph quickly mastered an array of one-leg yoga moves, trap bar dead lifts, and hip hinges. A year later, reportedly Steph could deadlift four hundred pounds. The training changed Steph's outlook and his actual look. Steph started to walk upright, to jab his feet into the ground, to jump at the rim with his shoulder in an opponent's chest.

With a few seconds left on the clock in Game 6 of the 2015 NBA Finals, Steph looked into his teammate's eyes, grinned, and yelled the word *WHAT?* Curry had some of the NBA's strongest hips, and the first of many NBA championships. Over a decade without ankle issues, he won it all in 2015, 2017, 2018, and 2022, while perfecting an array of little victory shimmies.

How could my hips win a championship? We're trying to fix problems that are largely invisible. I ask P3's Jon Flake if he can spot a good set of hips with the naked eye. "I mean," he pauses. "No. I haven't, like, seen someone walk in the gym and been like, 'They got a real good set of hips on him.'"

Hips carry a whiff of the forbidden. The babymaker, the moneymaker, the base chakra. This ball and that socket share a cul de sac with poop, pee, and procreation. Zero adults in my childhood discussed hips; trochanters and acetabulums didn't come up in high school biology. My basketball coach, who would happily go on about knees and ankles, slipped no hip words past his Americana lips. Down under the bathing suit, it's *personal*.

One evening, under a potted palm in the garden of a Santa Barbara microbrewery, I ask Marcus to dumb it way down. Say I wanted to build a hip out of Lego. What are the pieces?

"There's seventy, eighty, a hundred components," he says. "I can start listing them." He shifts tactics. "It's a socket joint that can do a whole lot. And it's actually hyper-complex. And mostly, they work pretty well, despite the fact that they're not cared for.

"But when they work really well, it's *just great*."

It's a stretch to say that hips made humans brilliant, successful, fast, verbal, and globally dominant—but not much of a stretch. Cheetahs run by reaching four legs through a poetically massive range of motion. As the back legs grab the grassy plain out in front of the shoulders and the front legs damn near reach the butt, the cheetah's torso curls into a croissant.

That all-in squeeze makes cheetahs a little faster than antelopes and way faster than us. But curling so completely demands empty lungs; cheetahs have no choice but to exhale with every top-speed step. It's borderline hyperventilation.

Sports scientists counted: Usain Bolt set world records taking one breath per twenty steps. Thanks to our hip-bifurcated bodies, our legs and lungs don't fuss with each other. Along with hair instead of fur and sweat instead of panting, free breathing is the reason we can run all day.

Running all day was, for a time, our essential evolutionary advantage. The other hominids that we outcompeted didn't have our running tools, like prominent glutes and Achilles tendons. When the parched springbok or giant sloth keeled over in the heat, humans got the high-density food that led to an alarming increase in our brain size.

In his book *First Steps: How Upright Walking Made Us Human*, Dartmouth professor Jeremy DeSilva adds another note: by being on all fours, carrying heavy weight in the shoulder, four-legged animals are structurally inhibited from making the fine-tuned sounds of language. The late biomechanics expert Robert McNeill Alexander wrote in the *Journal of Anatomy*, "I have compared the gaits of animals that walk or run bipedally, with human gaits. The general conclusion is that no animal walks or runs as we do."

In other words, hips let us be upright and explosive. Hips let us breathe better and run all day so we could hunt the protein-rich animals that fed an explosion in our brain size, facilitating the development of language. Somewhere in there, we became the only animals that make art and ponder our place in the cosmos. Hips made us brilliant.

Our hips' physical structure addresses our unique body shape: humans are top-heavy topple risks. Adult heads weigh more than ten pounds, your two arms together about the same. Half of your weight, or a bit more, is in a torso's worth of ribs, heart, liver, spine and gizzards. On average, less than 20 percent of your weight is your legs. Wrapped in a tangle of bulky and strong glutes, quads, and the body's toughest tendons, hips can clench like a fist into a heavy, wide, base to keep our torsos steady above our legs. But that's only some of what that complex joint can do. Those soft tissues can also relax, and flex into ballet—the body's most graceful, mobile, and powerful joint. Our hips can be soft or hard, rigid or mobile, strong or bendy—so much of healthy human movement is about getting the mix right.

About the age of fifteen, three big bones—the ilium, the pubis, and the ischium—fuse to make the body's one real socket, with the fun-to-say name *acetabulum*. The longest bone of the body is the femur, in the upper leg. It's so unruly that evolution deposited anchor points here and there for soft tissues to exert some control. The bony knobs at the outside of the front of our hips, the greater trochanters, are where muscles like the gluteus medius and piriformis grab hold.

From the hips' wide point, the femur turns and travels so far up your groin that the ball and socket approach the belly button. The femoral head is snuggled on all sides by a tough cuff called the *labrum*, which does a lot to hold the femur in place. As it has no nerve endings, a damaged labrum doesn't cause pain. (The pain comes from what follows: an unstable femur crashing around.)

So many muscles fold around, over, and through the hips that even doctors use shorthand. The vastus medialus, intermedius, lateralis, and rectus femoris are *quadriceps*. When the rectus femoris works with the iliacus, psoas, iliocapsularis, and sartorius, they're known together as *hip flexors*. The back of the hip stars three different muscles, or *glutes*.

The gluteus maximus does the heavy work. ("Athletics," explains P3 trainer Jack Anderson, "*is* hip extension." Your butt is back, then your hips snap forward, and you're off.) The glute medius and minimus push the leg laterally, out sideways. It's called *abduction* when you push a limb away from your body and adduction when you pull it back to center. The groin is the main leg adductor.

*Piriformis* means "pear-shaped," and that muscle runs from that dimple on your lower back to the knobby trochanter on the outside of your hip. Along the way, the piriformis pretty much fills your sciatic notch (home of sciatica), which is, not coincidentally, pear-shaped. In its travels, the piriformis has time to become one of the "deep six" external hip rotator muscles that look, in anatomical diagrams, like the fingers of a hand stretching across your butt beneath the glute max.

Almost every hip muscle is underappreciated, misunderstood, and unknown. Google "TFL" and the top search returns are "Tackle For Loss," "Tips For Life," and "Transport For London." Pity the wizardly *tensor fasciae latae* as it runs helpfully from the base of the spine, around the outside of the hip, and toward the knee. The TFL has an incredible task list, including raising your knee in front of you, pushing your leg out to the side, and returning blood to the heart. The TFL also touches areas of special concern in P3's data—the rotation of the tibia and the tilt of the pelvis.

The TFL neighbors the mysterious *quadratus lumborum*. Researchers know the "QL" attaches to the lumbar spine, the pelvis, and a rib. And then it does . . . something. "It is difficult," says a recent National Institutes of Health publication, "to identify precisely the actions that occur through the contraction of fibers," and "a cadaver study has raised many doubts about the action of QL."

P3 decodes each of these mysterious components, sorting which of those dozen-or-so tissues is rigid, weak, angry, or slack. A groin might need myofascial release; perhaps a glute med would benefit from weighted side planks; hip flexors often beg for pigeon pose. Hips are personal and personalized: an equation with a dozen variables has millions of possible answers.

Nevertheless, Marcus says, hip trouble tends to tumble into two categories. Essentially, every person who has ever been assessed at P3 has a primary target to work either on hip mobility *or* stability—seldom is it neither, occasionally it's both. Marcus thinks it's within your powers to divine which club you are in, and he believes that would be super-helpful. "If you can just understand if you need more mobility, or you need more stability, in your hips, at least you're starting to get in the game," he says.

"The most impressive athletes," says P3's Leah Borkan, "choose how to use their hips, instead of moving how their hips allow."

A core job of the hip is to keep the femur on track, over the knee and ankle. Picture a freight train, steaming across the flats, engine chugging, and every car nicely spaced. When someone pulls the brakes, the train compresses a bit. Brake harder and a car might angle a little to the right or left. Keep at it, and the whole thing could derail. Of the train cars that make our legs—foot, tibia, and femur—the femur is the wiggliest. Without stable hips to hold it in place, it's prone to leaping off track—with most of your body mass behind it.

There are many flavors of hip instability. The one that worries P3 the most is rotation of the long leg bones. We might stand next to each other and squat. Our butts might move backward the same amount, our knees might bend similarly, our backs might park at similar angles. The trains seem to be on the tracks. But if we were at P3, the SIMI Motion Capture System might see, in the motion of the little balls affixed to our legs, a femur that rolls internally. To imagine it, hold your forearms out in front of you, palms up, then rotate palms down. As some legs bend, femurs spin like forearms, inward toward the groin.

Internal femoral rotation has long been associated with injury. One study, from the Mayo Clinic, used "30 fresh-frozen cadaveric knees" in a mechanical impact simulator. People with strong stomachs attach a once-beloved knee to a machine that twists the ACL until it rips.

The concept, established by research and confirmed by orthopedists who consulted with P3, is that ACL tears commonly occur with a nearly straight knee, a rotating femur, and a knee tilting inward. Rip a drumstick off a chicken and you get the idea. In P3's mountain of NBA movement data, femoral rotation is one of the best predictors of catastrophic knee injury.

This is why P3 seems blasé about Q angles. When the knees dip toward one another, it's not great, but in P3's data the huge risks come

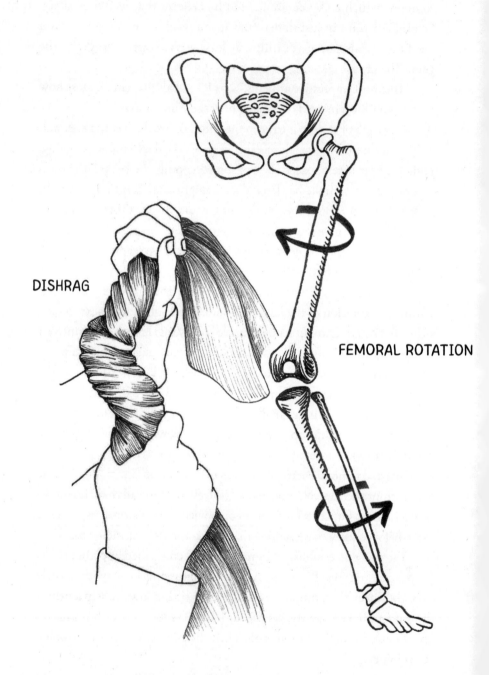

DISHRAG

FEMORAL ROTATION

not from the knees diving inward, but from the subset accompanied by a rotating femur.

Even harder on knees is when the tibia, the larger shinbone, also rotates. As you sit down in a chair, do your feet pivot a bit, so the heels come together and the toes end up pointing out? That's by far the most common way for a lower leg to rotate (unless, says Jon, "you're bow-legged like LeBron").

So, picture this: Some NBA players, as they land, have the upper leg bone, the femur, rotating in toward the midline of the body, while the tibia rotates outward, with the heels drifting together. The knee, then, is being wrung out like a dishrag.

I run this phraseology by Eric. He says P3 usually calls it *relative rotation*, but sure, we can consider it a dishrag. The point is that when the femur rotates in and the tibia rotates out, the issues are coming from the floor and the hip, but it's the knee that catches hell.

Eric calls translation and hip instability "the Tim Duncan and David Robinson of unfortunate knee injuries." Uninjured young athletes come in all the time, go through the assessment, feel great, kill the vertical jump, and are surprised to hear they are on the cusp of crisis—not unlike feeling healthy when a doctor says your blood test shows you're at risk of heart disease.

Marcus says his first yoga teacher moved her hips in awe-inspiring ways. "She would show off, you know? She did yoga six days a week." But her generous range of motion eventually led to derailment. "She had both her hips replaced when she was in her early fifties," says Marcus. If she had come into P3, Marcus says, they would have worked on "strength, and built some stability around her hips."

In P3 assessments, women are slightly more likely than men to lack hip stability. And in Marcus's observation, people drawn to yoga already have mobile hips—they're the ones who can slip gracefully into the lotus during the warm-up and the half-moon twenty minutes later. Marcus's solution: Weight lifters and yogis should swap workouts. The mobile hips would get more stable; the stable hips more mobile.

A lean, six-foot-four outside hitter, Kim Yeon-koung is the Michael Jordan of Korean volleyball, once clocked spiking a ball at sixty miles an hour. When she arrived at P3 in the spring of 2022, Kim was planning a return to the court at age thirty-four, after four knee surgeries, thumb issues, and an abdominal tear. The NBA's seven-foot, one-inch Chet Holmgren was at P3 at the time. Both look like they might snap, like twigs. Kim was at P3 to revitalize her career, chiefly by stabilizing her hips.

"Before you leave here," P3's Jon Flake told Kim, "you're going to do a one-hundred-kilo trap bar dead lift." That's the weight of a baby elephant, or a full-sized dolphin. Kim had been a professional athlete her entire adult life and had never deadlifted anything close. "She thought," Jon says, "I was insane."

There are two main ways to lift heavy weight: with a squat or a hinge. P3 uses both, but emphasizes the hip-stabilizing hinging movements. Careless heavy lifts, the pinnacle of bro culture, can create alarming injury risk. But coached carefully, and built up over time, research shows big weights teach a body to recruit ever more surrounding tissue to support the hip's movements. The dead lift builds many hip-stabilizing muscles, including glutes, hamstrings, quads, erector spinae, rectus abdominis, obliques, and an array of core muscles.

From above, a trap bar looks like a metal hexagon with pegs on two ends to hold weights. It lets you step into the middle, squat down, and grab upright handles just outside your shins. Research shows athletes can lift more weight, more explosively, with less injury risk with a trap bar than a traditional straight bar.

Nearly two months later, Kim's numbers had improved all around. In a positive sign, she had added two inches to her vertical jump. She felt and looked confident—until Jon announced it was time for the one hundred kilos. "Really? Now?" Kim quickly seemed unsure. She managed a ninety-kilo test. Jon barked, "NO PROBLEM!" Kim said she'd had a sleepless night. Kim's the kind of megastar who's trailed by a social media team. She told the camera, in Korean, that a hundred kilos was "a bit too much."

Kim circled the weight and paused. Then she stepped in and grabbed the handles. From her neck to her calves, muscles strained. "Get lower, get lower," hollered Jon. Kim reset, dropped her hips, brought her chest up. Her face tightened. NBA players, college athletes, the social media team—everyone stopped. All eyes on Kim. She stood a little. The bar climbed, past her knees and up. It seemed far slower than most gym moves. At last, she arrived militarily upright with shoulders back and glutes locked.

Kim sang a high note of elation. Jon yelled, "I TOLD YOU!!" Kim put the bar down. Greg Brown, a smiley Blazers forward, ran to congratulate Kim.

"That's it," Jon said. "Game over." Later that day, Kim would be airborne, home to Korea. Kim relaxed, leaned down, and hugged the shorter Jon. When they let go, the sound of *awww* filled the room. Was the hardest hitter in volleyball . . . crying?

Jon pointed up at Kim's face and said, "You don't cry!"

Kim corrected him, in English: "Women don't cry."

Back home, Kim led her team to the best record in the league, enjoying her best Korean league season in fifteen years. Reviewing highlights, P3 staffers noted her hip stability.

Hip immobility might be easier to understand than instability. Anyone who was once a kindergartener and can no longer sit crosslegged on the floor knows it. "I kinda feel like people who have tight hips," Marcus says, "know they have tight hips." You know it from your efforts to touch your toes, to get on and off the floor, to remove a splinter from the sole of your own foot, to squat without lifting your heels off the floor, to do yoga. In an assessment, P3 staffers see it by measuring the femur's movement in every direction—up toward your chest, across your body, to each side, and toward your butt. With a bent knee, they swing your shin in and out, testing internal and external rotation.

Despite the freight train analogy, a rigid hip is not perfect, because in fact humans don't move on tracks. We rotate, lunge, lean, and flex. Stiff hips tend to disinvite some of the body's most useful soft tissues from the landing party. Blenders, for example, have hips that don't move backward on landing. The glutes don't engage, the force is passed into the lower back. In research, poor hip mobility is associated with an array of knee issues and back pain.

Marcus has spent a lot of time with Aaron Gordon, one of the NBA's most breathtaking athletes. Aaron's older brother Drew was once the starting center for UCLA and a top prospect. Drew showed up in

SIDE PLANK               FIGURE FOUR

P3 data as just a tiny bit stiffer than Aaron in the ankles and hips, "and ended up," Marcus says, "with a big knee injury and had some chronic changes to his knee. He had a back injury because he doesn't—when he lands, he doesn't take much load through his hips." Before Drew's tragic death in a 2024 car accident, he mostly played outside the NBA, in summer league, the G League, and overseas.

Drew and Aaron arrived with similar assessments. "It was actually kind of an amazing look," Marcus says. "It felt like it was almost the same person." After seeing his brother's career plateau, Aaron spent years working to improve his own hip mobility. "He has enough mobility," Marcus says, "right at the borderline of what he needs to be able to execute these movements."

Aaron has excelled in the NBA for more than a decade, won a championship, and earned more than a hundred million dollars. But Marcus says that when he watches Aaron jump, he worries a little. "I wish he got a little bit more out of his hips. I think that his knees will probably be an issue as he gets deeper into his career," Marcus says.

Marcus has shortcuts to assess your hips' needs. "Everybody knows what a side plank looks like," he says. Marcus likes to see people do a tough variation with arms and legs straight, all ten toes pointed forward, and the upper leg raised. "If people can lock out, especially if they look like an *X* when they do that—because their legs are perfectly stable, arms perfectly stable, and they're just locked in that *X* position—I feel like that checks the box" for hip strength.

And how, I ask, can people tell if they have immobile hips? Get into the pigeon position, or a favorite featured in the standard P3 warm-up: the standing figure-four. (Stand on one leg, put the other ankle across the top of the knee, and then sit down, pushing the butt back as far as possible.) That won't test hip mobility in every direction, but it does test for the dead-common issue of poor hip external rotation, which Marcus says is both important itself, and a decent marker for general hip mobility. "If you can't do those," Marcus says, "then you probably need to work on your hip mobility."

Good hips can set you free. When Kevin Durant visited the bench press station at the 2007 NBA draft combine, he was a sure thing. He had scored like water at Texas and was nearly six foot ten. But when he failed to lift the 185-pound bar a single time, a Greek chorus of sports commentators insisted Durant would need thirty or a hundred new pounds of muscle to fend off the NBA's beefy LeBrons.

Maybe the only person who disagreed was Justin Zormelo, whom Durant hired to prepare him for life as a pro. "I just felt LeBron couldn't chase him," remembers Zormelo. "Why do we want to tone a guy of his level down, when we have a competitive advantage?"

Durant danced away from defenders and in time became a champion and an MVP. His tough-to-predict movements emanated in no small part from his hips. In the past, big men had relied on height alone. They were typically poor movers. Fifteen years after Durant's arrival, Zormelo says, "the game has changed." The league is enthralled at the prospects of Durant-or-taller players like Victor Wembanyama and Chet Holmgren who are also light and fluid. Zormelo starts every workout session with hip mobility work, because, he says, basketball is becoming soccer, soccer is about freedom of movement, and freedom of movement is about hips.

In a 2022 article in the academic journal *Sports*, the P3 staff notes that "basketball players spend approximately 31 percent of game actions shuffling laterally." That demands copious hip training—side planks, hip dips, bridges—which, it struck me, looked more than a little like the work my mother-in-law did to rehab after hip replacement.

Does it seem a little like the work NBA players do to move laterally might also work to shore up the brittle hips of the aging? Eric bellows, "I THINK SO!!" When Paul Dudley White enlightened the world about the cause of heart attacks, he suggested that the solution might be exercise, diet, managing stress, or reducing smoking. He prescribed those things to the president. But he added that he had no definitive answers without more research. To Eric, sports' worst injuries are having a similar moment. The cause is coming into focus. The research into best remedies will take time.

In the meantime, it would make sense, Eric explains. Professional athletes encounter the same movement challenges we do, but with more force and frequency. In some ways, they simply age faster, which puts some of them on the express train to hip degradation. What seems to work for them is taking command of the posterior chain, especially glutes and hamstrings, to get better control of the femur's movements. Until there's peer-reviewed research to confirm it, Eric says that at P3 they "tend to assume that's two birds with one stone." It'll get you moving, and keep you healthy.

Total hip replacement surgery has become one of modern medicine's most ferocious growth areas. A 2023 *Journal of Rheumatology* article predicts there will soon be more than a million a year in the United States alone. At the current average of almost $40,000 each, that's $40 billion a year.

From Jon's point of view, there's much more we could all do to keep our hips from wearing out. Did your hip wear out because you used it too much, or "because you didn't use it? You lost it? Was it years of sitting behind the desk?" Jon thinks about all the hips that don't really move.

The stair machine comes in for particular criticism. As a basketball player at Princeton, Bill Bradley moved so stunningly well that John McPhee wrote a whole book about him called *A Sense of Where You Are*. Bradley won championships in the Ivy League, EuroLeague, Olympics, and NBA. Then he continued his extraordinary career in the US Senate. Bradley wrote about the hard work he put into the Senate gym's stair climber—which he blamed for the degradation of his hips, both of which have been replaced.

Hip rejuvenation is not enjoying the same explosive popularity as hip replacement—nor is it covered by insurance. "It's expensive," Jon points out, "and takes time." Surgery comes with its own hard work, pain, and physical therapy. But Jon, who teaches movement, notes that at its core, "it's a passive thing, isn't it? Yeah, you lay down, go to sleep, and you wake up with a new hip."

The work of keeping a hip functioning, Jon says, "is the opposite of that." Athletes sweat to do it every day at P3. NBA player Mikal Bridges has some of the best hip measurements in P3's database and is the NBA's most robust player, having not missed a game in over five years. Who wouldn't want to copy him? Chet Holmgren spent his predraft months at P3 exploding laterally up and down the track, lunging sideways with a kettlebell in his hands, learning to snap his hips forward explosively, or dragging a coach across the floor using a band wrapped around his waist. Over seven weeks of predraft training, Holmgren improved his lateral force production almost 11 percent, and his change of direction 7 percent.

We could fill a book with P3's hip work. A snapshot from a single day:

A teenage boy puts black booties over his shoes, then bursts right and left in a skater motion across a smooth surface called an UltraSlide.

Almost everyone takes a turn putting a piece of padding foam across their lap, while pressing their back into a red-and-black device called a Sorinex hip thruster. With feet on the floor and a heavy barbell across the pelvis, the task is to press the weight into the sky.

A soccer player in slippery little booties lies on the floor on his back, with hips elevated, and presses his heels out until his legs are straight, and then retracts them.

A common glute exercise is to put a band around the waist, affixed to something solid behind you, and then to push your hips forward into the band. P3 adds a wrinkle: a two-footed standing broad jump into the band's resistance.

Lying on her back, a soccer player lifts her hips off the floor in bridge pose—but with a heavy medicine ball squeezed between the knees.

A strong-looking teenager tows a heavy sled backward down the track from a strap around his waist.

You can see the results. There are many shapes of athletes at P3— German soccer players, Canadian NBA players, Texan NCAA volleyball players. There's almost nothing that every P3 athlete has in common except, to my crude eyes, glutes. It appears impossible to escape a course of P3 training without building strength where it matters most. P3 spends a lot of time teaching athletes how to bring strength and mobility to the glutes, the hip flexors, the piriformis, and the big muscle that's secretly at the center of everything: the psoas.

# 15

. . . . . . . . . . . . . . . . . . .

## PSOAS

. . . . . . . . . . . . . . . . . . .

*Where you think it is, it isn't.*

—Ida Rolf

ON MY FIRST trip to Santa Barbara, Marcus walked me to P3's sister business, the Lab, which is sometimes referred to as "P3 for normies." There I met Mike Swan, the physical therapist who rented Marcus the floor space where P3 was born, and who has earned a reputation as a high priest of soft tissue. (I once watched a woman, who had been scheduled for back surgery, undergo one session with Mike, then stand up and say she felt all better.) About four minutes after meeting me, Mike told a very unsurprised-looking Marcus that I lacked hip mobility.

Marcus nodded, smiled, said, "Good luck," and walked out.

At that moment, I would have described myself as an atheist, unsure there was such a thing as a human soul. Then Mike jammed eight clarifying fingertips into the front of my right hip: *Yes, I do have a soul, and Mike is choking it.* My heart rate spiked, sweat gathered at my temples, my sense of humor crumbled. After thirty seconds, when I thought he might wrap it up, Mike took a wistful glance at the clock and lamented that we only had twenty minutes left.

That, Mike said, was my psoas.

"Huh," I said, between labor-and-delivery exhales. "I thought the psoas was in the back."

"Oh, it runs back there," Mike said. Then he demonstrated its shape by fluttering his hands like *My Octopus Teacher*.

As Mike spoke, I was decades into caring deeply about athletic bodies. I knew about visible muscles like biceps and quads, but had mostly ignored the invisible ones that keep us running. Studies have correlated a weak psoas with bone wasting, surgical complications, poor prognoses in cancer treatment, and—amazingly, and for reasons that aren't well understood—death.

The psoas works hard. It's the only significant muscle that exists both in the top and bottom halves of the body, and also the only muscle that truly moves around the body's center of gravity, where it stabilizes both the lower spine and the head of the femur. It's the most important of all the hip flexors, which are the muscles that pull your leg up in front of you, as if to hurdle. It's integral to posture, walking, running, lifting your body off the ground, and moving laterally. It's also connected to the neighboring abdominal muscles and diaphragm, meaning the psoas has a deep back-and-forth with respiration.

PSOAS

Somehow, that's not all. The psoas is latched tightly and intricately to the discs, vertebrae, nerves, and blood supply of the back's bottom six vertebrae. The nervous system manages the entire lower body through an intricate web of nerves called the lumbar plexus, which is embedded through the psoas. The psoas looks, on an MRI, like a nerve center weaving through the kidney's intricate systems, the diaphragm, the esophagus, the intestines, the sex organs, and the pelvis's many wonders. The psoas might be the body's deep state.

Modern MRIs give researchers highly accurate ways

to measure muscle volume. The deltoids at the top of the shoulder, for instance, average 380 cubic centimeters. The triceps are almost as big, followed by the pectorals on our chests, which are typically 290 cubic centimeters.

The psoas clocks in with an average volume of 407 cubic centimeters. It's huge, and most of us have no idea where it is.

Lie on your back and find the knobby bone on the front of your hip. Now let your fingertips walk a tiny step into the soft tissue in the bowl of your pelvis, and press down. If you're in the right spot, as you lift a foot slightly off the ground, you'll feel something like a bass string pressing into your fingertips. That's your psoas saying hello.

From under your finger, the psoas travels to interesting places. Mostly, it plunges through the innards behind your belly button to grab the inside of your lower back. When you sit up straight, the psoas pulls your lower back forward, from the inside.

At the hip, the psoas joins forces with the *iliacus*, a muscle emanating not from the inside of the spine but from the inside of the pelvis. Together they have much to say about your posture, gait, and movement.

The *iliopsoas* also travels to a more secluded spot. If you're curious, alone, and not squeamish, put your hand in your groin muscle, and dig around until you find the femur beneath. Follow the femur up, and up a little more. Near the top, there's a knob. When your fingers find it, you'll be in a prizewinning crotch grab. *That's* where the iliopsoas attaches.

This muscle is central to our multifaceted disability crisis. "An estimated 619 million people," says the World Health Organization, "live with lower-back pain and it is the leading cause of disability worldwide." A growing body of research shows that those people tend to have measurable atrophy or weakening of hip muscles like the psoas. Unsurprisingly, the psoas tends to get weak and stiff from sitting—and sometimes angry from running, biking, or swimming.

Psoas workouts are a hornet's nest of theory and opinion. In video footage, mixed martial arts fighters train by scrambling around on all fours like wild creatures, heavy with psoas-rich movements like pulling your knee close to your chest. (We evolved cooking, eating, and socializing in positions like that. Perhaps psoas workouts are a return to normal.) Impassioned essays implore you to stretch your psoas; other impassioned essays implore you not to. Even at P3, I've heard trainers saying that stretching a psoas sometimes angers the thing.

There's broad agreement that a strong psoas tends to be happier than a weak one. While at P3, Mexican professional soccer player Juanjo Purata does a lot of exercises with a "march" component, which means pulling your knee up proud and high—using your psoas—as you stand on one leg. In one variation, Purata marched up and down the track with a heavy kettlebell at his side like a suitcase and a soft medicine ball held aloft in the palm of his other hand. Later, he chest-passed a medicine ball to himself off the wall, again and again, with a leg up high in the march position. At P3 they also find that, after an array of hip work, clients tend to be able to raise their knees higher toward their chest.

Glute work also seems to help the psoas. Many of P3's exercises work with the agonist/antagonist principle—it is well established that some of the body's muscles are neurologically paired so that as one contracts, the other relaxes. That's why a yoga teacher might say to tighten your quad while stretching your hamstring. The psoas operates largely opposite the glutes, which is why Purata paired his march movements with glute work. He sank his hips back into a squat with resistance cables running from each hand to the Keiser machine, and then in one exciting thrust, he pulled the cables to his chest as he snapped his hips straight. He did Romanian dead lifts on one leg, but instead of lifting a weight from the floor, he pulled a cable from the Keiser machine in front of him. The wide world of psoas therapy looks a heck of a lot like all the work P3 does to develop the posterior chain. Squats, step-ups, jumping forward against a band, pushing a sled, hill sprints—everything that gets your hips strong, extending, moving with range, attenuating force—tends to also help your psoas.

When elite five-kilometer runners added hip flexor exercises to their routines, MRIs found that increases in psoas size correlated nicely with faster times.

A simple way to check in with your psoas: Stand on your left leg and raise your right knee as high as it'll go under its own steam. If you sit a lot, it might stop parallel to the floor. If the P3 coaches saw that, they would quietly make plans to get it higher.

With your leg up, test your hip's rotation. Picture your knee pointing at the center of an imaginary clock on the wall in front of you, while your right foot dangles at thirty minutes past the hour. Keeping your hips straight, sweep the foot clockwise as far as it will go, and then counterclockwise. Experts suggest that between those two moves, you ought to be able to sweep fifteen minutes, commonly from twenty to thirty-five minutes past the hour. If you can't, you might have a bony limitation, weak muscles, or the need for myofascial work.

Marcus doesn't drink much caffeine—typically a single morning cup of green tea. But one afternoon, he started messing with P3's espresso machine. By the time Eric saved him, he had commingled decaf beans with regular. (As he fished the decaf back out, Marcus asked, to no one in particular, "Why do we even have that?")

After a strong espresso, Marcus talked fast about the perils of being tight in the front of the hips. He had just met a low-ranked but high-hoped professional tennis player. He tested as an exceptional athlete—six feet, four inches, with long arms, 205 pounds of lean muscle—except for one thing.

"Laterally," Marcus says, "he's terrible. He can't create force out of his hips." The player's anterior hips were so tight that the strong muscles that would move him side to side were essentially unavailable. A number that matters to Marcus is newtons per kilogram—the force you generate, divided by your weight. In an embarrassed whisper, Marcus disclosed that this player generated a mere 7.4 newtons per kilogram in the lateral skater. This guy's trying to support himself playing tennis, has a coach, and works out constantly, Marcus explains—and laterally, he generates the newtons of a middle-schooler. "He's got a peashooter for hip extension," says Marcus.

"We've done in-house research on mobility," Jon says. The news is mostly good. "And what we have found is that anterior hip mobility is very plastic; it can change quite easily."

"Your body," says Marcus, "is crazy adaptable." One of the ways P3 would adapt that player is with the move that made me consider the divine: the psoas release. Peer-reviewed research has been so-so on the long-term effects of myofascial work. A recent survey found the available studies "mixed in both quality and results," but ultimately "encouraging." But in Marcus's view, well-applied myofascial release can make the difference between making it in professional sports or not.

That might be surprising if you're used to thinking of a body through the lens of bone bias. X-rays have been around for a long time; bones are clearly diagrammed, stable, conceptually like the two-by-fours that frame our houses. "Did you break it," they ask, "or just sprain it?" The implication is that real injuries happen to masculine and well-defined bones, while soft tissue seems more feminine and mysterious. Meanwhile, our upper and lower halves are held together only by ligaments, tendons, and psoas.

The psoas release is a minor miracle. For all its grimy emotions, the aftermath comes with the divine feeling that your hips have returned to their rightful place. Under a YouTube video showing rudimentary psoas maintenance, commenters find they need all-caps: "LIFE CHANGING. This was the stretch missing in my life!!! THANK YOU," says one. "No lie after about 5 minutes I feel SIGNIFICANT relief from a problem that has literally plagued me for years," says another, while a third notes, "I simply CANNOT overstate how amazing the difference has been."

We are all Marcus's tennis player, lacking this aspect of hip enlightenment. "He needs to develop some self-care," Marcus says. "He's got to work on mobilizing, stretching all the anterior hip muscles, all of his quads," says Marcus. "And he can get a thumb dropped on him at least once a week." It upset Marcus that the player had gone twenty-four years, and played four years of high-level NCAA tennis, but still awaited his first psoas release. "Like, how does that not happen?"

You can release your own psoas. The Lab teaches first-timers to use a hard medicine ball, about the size of a volleyball. I bought one at Walmart for twelve dollars. Get in the plank position, press the psoas on the front of your hip into the ball, and then sink your weight into it. The Lab's take-home instructions specify "raise leg to dig in further." When you find the right spot, it feels wrong.

Or, lie on your back with a lacrosse ball on your psoas, and then rest a heavy dumbbell on the ball. Some people prefer the twenty-three-dollar psoas release device shaped a bit like a cane with a hard ball affixed to one end. Lie on your back, jam the rounded end into the front of your hip, and it'll give you a sense of meeting Mike. I like to apply the pressure where Mike did, just inside the pointy hip bone, but others prefer an inch or two higher, toward the belly button.

One early morning, I joined Eric for an employee-hour workout. I asked if P3 had a psoas stick I could use. Eric handed me seven feet of dense wood, a wizard's staff with a ball affixed to one end.

As Eric resumed his workout, I lay down on my back and jammed that massive tool in place with two hands. When Eric caught sight of my little scene, his face erupted in delight. "What," he asked, "are you doing?"

Only then did I grasp the absurdity of waggling a seven-foot wizard's staff from your crotch halfway to the ceiling. Eric took the thing, walked to the wall, and calmly demonstrated how it's actually used. Stand up, plant one end of the stick in the crease where the floor meets the wall, and lean your psoas into the ball on the other end. Incredible.

P3 tries to help athletes find good soft-tissue workers all over the world. It's mostly trial and error, but the CEO of the Lab, Alex Ash, says he looks first for recommended "orthopedic physical therapists," and failing that, Rolfers. They have no interest in spa-type comfort. This is about joint function.

One of the Congo's best-ever basketball prospects is six-foot, eleven-inch Tichyque Musaka. When he first came to P3 in the summer of 2022, he raised eyebrows with an incredibly low-tech hip mobility measure: He lies on the physical therapist's table, face down, while Eric pushes his heel toward his butt. At a point that Eric swears is precise, he stops pushing, and measures the distance, in inches, from butt to heel. The "heel test" is the most kindergarten-level datum in P3's futuristic array, and important. Big changes in this test, over time, can signify serious stuff.

Many athletes can touch heel to cheek, but not Tichyque. On first assessment, he could only get his right heel within seven inches. Eric wrote 7 on the clipboard. After further assessment and some discussion, the team focused on the quad, which evidently had a death grip on Tichyque's femur.

Some immobility comes from muscles being a little clingy, like a child scared to let go of a parent's hand. Sit cross-legged on the floor. If your knees poke up like the hills above Santa Barbara, often it's because your groin muscles won't let the femur go.

Active release—essentially pressing with fingers—can quickly help. But, in keeping with the hip's reputation as a cultural hot zone, the key spot where the adductors grab the femur is at the tippy-top of the groin. When that muscle is seized, P3 teaches athletes to release it themselves. ("Nobody," Jon points out, "is trying to get canceled.")

"So, basically," Jon explains, "you put your actual adductor on the metal handle of the kettlebell and you rotate, flex, and extend. . . . Most people are like, 'Oh, my God, this feels horrible.'"

"Even though research is a little split on foam rolling and self myofascial release . . . in my eyes, from what I've seen here, and what I've done on myself, I think it works pretty well," says Jack. "It creates a window to enhance the ability of the stretch we're going to do next."

This is how Tichyque met the quad smash. He sat on the floor with his back pressed to the cinder-block wall, legs straight out in front, as Jon fetched one of the heaviest single items in the gym, a fire truck–red thirty-two-kilogram kettlebell.

Tichyque parked those seventy-odd pounds of cast iron directly on the quadriceps. Quickly, the upbeat demeanor drained from his face. He muttered something to a passing Jon, whose retort was beyond sunny. "BREATHE, BREATHE!" Jon hollered, projecting the opposite of concern. "IT'S GOING TO BE OKAY!!" The first round lasted two minutes.

Later, Jack said that professional athletes generally take this move in stride, but the first time teenagers quad smash, "it's like they're being shot." Tichyque was eighteen.

After Tichyque switched the kettlebell to the other leg, Jack strayed near, and Tichyque spoke through clenched teeth. Jack grinned and gave him a high five.

When that was finally over, Jack and Jon brought Tichyque to a rig set up with the strongest band in the building, a loop of rubber perhaps eight inches across and six feet long. The teenager knelt before the rig like Vader before the emperor, one knee on a pad, the other sole flat on the floor, with the stretched green rubber looped behind his kneeling thigh. The band pulled Tichyque's femur forward with such force that Jon and Jack held his upper body in place, to keep him from shooting forward. Sweat trickled down Tichyque's anguished brow.

Like grief, the panic of myofascial release ebbs with time. Every P3 workout features soft-tissue work targeted to the day's workload. One day's workout might prescribe a minute each side, so eight minutes total, putting a lacrosse ball into the psoas, glute, TFL, and QL. Another day might pair a hamstring release on the med ball with a banded stretch. All that poking into your muscles can be tense, but over time it comes to feel more like an oil change. Things run smoother. And then, at some point, you get a sense of why Marcus seems to age differently, and better, than most of us. He doesn't wake up mystified by some fresh hell that has seized his hip. He wakes up thinking, *Huh, my psoas is a little tight*, and then he addresses it, and it's better.

# 16

. . . . . . . . . . . . . . . . . .

## RELAX

. . . . . . . . . . . . . . . . . .

*Thinking, for us, was a disease.*

—Lonnie Jordan of the band War

WHEN NADINE LIED to Marcus about what time the plane took off, he was, she remembers, "so mad."

"I'm still mad," Marcus says, a decade later, "just talking about it."

Nadine says she would happily get to the airport three hours early. After a pause, she adds, "I'm German."

Marcus, though, puts real thought into damn near missing every flight. Just as target shooters account for wind and heartbeats, Marcus juggles traffic, parking, and security lines so that he can walk up to the gate and show his boarding pass in the sliver of time after everyone is seated but before they lock the jetway door. Once, he says, he took his seat without breaking stride from the car, and it felt "perfect."

When he added that he had never missed a flight, Nadine made a face. Once, Marcus admits, there was "unforeseeable" construction between Santa Barbara and LAX. Nadine nods and adds, "Exactly."

I asked Keean how he felt about arriving at airports. He grew quiet. The issue landed him between parents. Keean noted that his family has "definitely" been called over the airport PA system. But they have not missed a flight. "It always seems to work out," Keean says.

Nadine sees two things happening. The first is that 99 percent of the time, Marcus "has thought through everything perfectly." The other is that he seems to have an inability to experience normal anxiety.

Nadine has a friend who says Marcus is charmed. "It only takes one bad experience," Nadine says. If someone stole his wallet one time, he'd be careful every time. But instead, he leaves it all over, says it'll come back to him . . . and, to Nadine's amazement, it does.

This seems like just another "ain't Marcus quirky" story—nothing to do with injury prevention or elite athletic performance. But Marcus says he sees anxiety hindering free movement all the time. Relaxation matters deeply to a good jump, landing, and all ballistic movement. It's widely accepted as a core principle of elite sprinting. (Running guru Steve Magness posts, "Fast and relaxed is the name of the game.") Relaxation is a key takeaway of the misogi, which teaches the importance of avoiding an energy-sapping panic.

More specifically, anticipation fuels biomechanical problems. Some blenders are so eager to jump that they fire the hips before the ankles and knees have finished landing! Watch the slow-motion video and all you can think is, *Relax*. Everyone knows you can shoot a rubber band farther if you pull it back more—let it go early and it'll bloop limply to the floor. Our rubbery muscles work similarly. A sense of urgency is usually seen as a good thing, but it can mess with timing.

Marcus dedicates a lot of effort to his health—arduous workouts to stabilize and mobilize his hips, monitoring the range of motion of his surgically repaired knee, weekly yoga, surfing, annual misogis, and far-flung adventures. But his most pervasive project seems to be training his mind on the here and now. He might be the least anxious or distracted person I have ever met. People toss around the term *flow state*, which may apply to the way Marcus brings his whole attention and focus with him everywhere. All day, his full attention is on the athlete standing before him in P3, the breeze through the canyon, or Mila's art project—and almost never his phone.

Related: Marcus hates packing. "Too much anticipating," he says, "definitely fucks you up. It just gets you out of living." Nadine rolls her eyes at her husband's "bad planning," as he hikes through summer heat without water and freezes in winter without gloves. Not planning, though, is part of the plan: he trusts that he will cope. "That's just who I am: bet on your adaptability."

Marcus says he felt terrible for his medical school classmates, almost all of whom, he says, studied more. They carried the heavy freight of short sleep, overwork, and stress, as Marcus swam in the Brookline Reservoir, biked everywhere, and held raucous weekly game nights.

The first time I ever visited P3, Marcus introduced me to a robust curly-haired guy named Alex Ash. We ducked into a quiet back room, to do, Marcus said, "a little breath thing" inspired by a visit from Stanford resident and free diver Robert Lee.

Marcus and I lay on exercise mats on the wooden floor. Alex turned off the lights and asked how long I thought I could hold my breath. Maybe a minute? (Later, I'd find that most people can hold their breath for thirty to ninety seconds.) Alex said, "Great, let's try that." After I made it almost a minute and a half on Alex's watch, he asked why I had taken a breath.

Did I really have to answer that? We agreed that the reason was because I felt a little panicky. The fear of suffocation is one of humanity's most ancient and powerful monsters, and the reason waterboarding is torture. Not inhaling makes it simple to imagine death.

"Okay," said Alex. "The common story that we tell ourselves is that we're running out of oxygen, and so there's a lot of insecurity in that story, right?"

Absolutely. Alex said big-wave surfers (who can get pinned underwater by ocean turbulence) prepare by having a "friend" hold them down in the waves. That drill, in fact, helped inspire the misogi of running across the ocean floor carrying a rock.

"People who get really good at holding their breath," Alex says, "they're able to hold for seven minutes or more—and not have any damage. The urge to breathe, it comes more from a buildup of carbon dioxide than it does from a lack of oxygen." The ways of the lungs are complex, the neural response to elevated $CO_2$ is poorly understood. But there's plenty of science to support what Marcus says next: that the urgent feeling of having to breathe is a spasm of the diaphragm, the squid-like puff of muscle that causes your lungs to open. The feeling that creeps across your chest is like a pregnancy contraction, a cyclical tightness. Free divers understand that the contraction is negotiable, just as a runner need not immediately yield to the feeling of tired muscles. "As you get better and better," Alex said, "it's learning how to not just push past it, but *relax* past."

Alex offered to put a pulse oximeter on my finger, which would show that oxygen levels remain stable longer than you'd imagine. Like monsters, the feeling that we'll die if we don't breathe after ninety seconds also comes from the imagination. Alex said that if I held my breath long enough to cause worry, he'd stop me.

Then he led us through a series of aggressive exhales, to clear carbon dioxide, and instructed me that at the key moment, I'd fully inhale, lock the air into my lungs at the throat, and give a thumbs-up. Then he'd start the clock.

"You're going to start getting some of those similar sensations like that first breath hold," Alex said thirty seconds in. "And the key is going to be finding a way to relax through that. Relax through the contractions, see if you can even slow them down. Come up with some sort of story, right? . . . Some people will imagine a flickering flame. Other people imagine themselves just kind of floating in water."

Outside our quiet room was a lively gym. Now and again, you could make out a hollered word or two. Mostly, we floated in inky silence. The promised wave of chest tightness arrived about ninety seconds in. I thought about the months before my daughter, Molly, was born, months of urgent midwife and hospital visits owing to premature pregnancy contractions. They'd wheel over a machine called a tocodynamometer, and strap sensors to Jessica's belly. A wild contraction would arrive like a weather system. Jessica would grimace as her whole pregnant belly stiffened like a flexed muscle. But then, time after time, the storm would pass. The long spool of paper emitting from the "toco" would show a sine curve, the pressure coming and going like the tide, in literal waves.

The contractions I felt holding my breath came from similar muscles on a similar schedule. They might pass, too. If the monster of suffocation isn't real, then all I was doing was lying on a squishy mat in the dark; the easiest thing in the world.

I pictured a flickering candle. As promised, the first contraction receded, which felt immensely empowering.

"Two minutes," said Alex, ever so calmly. "Start trying to dive into a little imagery. Something that makes your whole body relax." The next contraction came a half minute or so later. But at some point, my mindset shifted from blissful flickering candle to grasping determination. My heart rate leapt. Alex would later tell me that they learned it's a big mistake to tell professional athletes that many people can hold their breath for three minutes, because then they try to win the breath hold with a competitive fighter's mentality, which dooms them. As the emotions swelled, I knew I was cooked. I slurped at the air. Alex whispered that it had been three minutes and ten seconds. I stayed in place, eyes closed, with Marcus somewhere off in the dark. His big breath came smoothly, through pursed lips, exactly one minute later.

I babbled about how much I loved the flickering candle, in a way that clarified who in the room drank the most coffee. "Let's do one more," Alex said.

Alex guided us through a series of ten-second exhales, followed by a raucous series of strong purge breaths, and then another minute of ten-second exhales, which turned into a thumbs-up.

Welcome back, little flickering candle buddy. At some point, Alex said, "Stay focused on the light." Aye-aye, Captain. After a messy approximation of relaxing through the first contraction, Alex said, "Two minutes." Then silence, more silence, more candle.

"Three minutes here," Alex said. "If contractions start to come, see how far you can allow yourself to push in.

"Feel that nervous system dropping down."

For whatever reason, as Alex said that, I perfectly defied the order and perked my nervous system up with anticipation: I traveled thousands of miles away and a few days forward in my head, home to New Jersey, where it would be fun as hell to tell my son, Duncan, how long I had held my breath. He loves that kind of stuff. That thought served up a Ping-Pong game of caffeinated thoughts, my heart rate climbed, and I lost the candle.

"Four minutes," Alex said as my physical sensation prickled with urgency. The ending exhale is a big *whoosh*. "Four-ten," said Alex calmly. "A minute longer."

In the dark, not breathing, Marcus was beyond still. (When I try this at home, my dog often wanders over to check for signs of life.)

Marcus exhaled at four minutes and forty-two seconds, a personal record.

The feeling afterward is like the calm, happy end of yoga class. "You did something that wasn't like jumping high," says Marcus. "Something deep inside me was able to find this thing. That's powerful.

"It's so essential to life, and almost all of it happens unconsciously," Marcus says. "We know you can affect your parasympathetic nervous system a ton by focusing on these really long, controlled exhales. You relax your nervous system."

A lot of P3 training assumes, as Marcus says, "your body already knows." I didn't teach my lungs and bloodstream to survive without air. Instead, I roused some dormant know-how.

Since P3 was a plyometric mat in the back of a PT's office, Marcus has been stingy about adding equipment. The stated reason is that he wants to preserve open space for athletes to move. (P3 staffers also suspect Marcus wants to preserve room to unfold the Ping-Pong table.) The equipment that makes the cut is a mishmash of conventional (weight-lifting rigs, free weights, Keiser cable machines, an assault bike), and niche (slideboard, kBox flywheel machines, a Tonal, a sloped track). And three custom-built plywood devices nobody recognizes: the *impulse boxes*. Alligator-sized half-pipes, they're pancake-flat across the bottom, angled forty-five degrees on each end, and perfectly designed to expose anxiety.

The boxes were inspired by Marcus's friend Brent McFarlane. Author of *The Science of Hurdling and Speed* and more than five hundred journal articles, McFarlane had his athletes almost tap dance. They'd start slowly and then build up to superfast paces on a twelve-inch plyometric box or a wedge-shaped device.

Watching McFarlane's sessions, Marcus remembers thinking, *This makes sense*. He has been teaching it ever since.

Adam showed me a video of former UCLA running back Christian Ramirez on the impulse box. Ramirez stands on the flat part in the middle. He jumps and lands on the ball of his left foot, pops up, and taps his right foot on the plywood slope out to his right, then lands again on his left. Then he switches feet, with his right foot in the center and his left foot popping to the berm on the other side. The contacts occur with the sizzle and intensity of a rockin' drum solo, a blur of triplets at a rate beyond counting. His face and trunk are still, calm, and happy.

There's a reason the box has sides. On the flat floor, the exercise would cause the foot to meet the ground at a steep angle, testing ankle and knee stability. The angle of the impulse box lets a foot, thrown out to one side, hit straight and stiff—starting a natural conversation between the hips and the feet without other joints making it weird.

There's a variant with a hard wooden wedge pressed into the wall. When he's at P3, Luka Dončić uses it, dancing in "feet of fire" with one foot up on the wedge and the other on the floor. After a few staccato seconds, he explodes laterally away down the track.

There's a lot of energy to be had from the body's natural elasticity. But to get that spring, you have to relax the rubber bands like your glutes—otherwise, Marcus says, "you don't get the freebies."

The stretch-shortening cycle has long been a hot-button frontier of biomechanics: how to get athletes down through eccentric landing and up into concentric explosion in the best way? "High-class athletes have been found to have a rapid alternation of these agonist and antagonist muscles," says the original P3 website, "due to their enhanced neural functioning from training. This process is known as intermuscular coordination." P3's workouts use "contrast complexes, which alternate agonist and antagonist movements to alter the timing of this braking burst." The trick is to train the brain to toggle faster from braking to exploding.

The "fire, relax" pattern can also be seen as "work, don't work." By the time athletes make it to the peak of their sport, and visit P3, they almost always know how to work. This high-achieving group most often struggles with the second command. Harvard students are often good at studying, worse at not studying. But when you're at the highest level and want to improve, Marcus says relaxation is often the answer. "There are always these additional levels that you can reach for through faster, more efficient relaxation," says Marcus.

In the video, Ramirez looks like he's dancing. Baseball player Cedric Mullins looks smiley and smooth while he impulse boxes at a thrilling cadence. Aaron Gordon looks fluid. NBA big man Andre Drummond is almost as fast as Gordon, while holding a medicine ball, still and quiet in his top half as he dances his six-foot-eleven way across the wood.

By the time Marcus coached me through the impulse box, I understood that the road to failure was paved in tension. I'd seen high-schoolers red-faced and intense on the box, like they're in a VO2 max drill, trying to crank up their speed. I was determined not to fail the test.

Marcus did a short demonstration, then I mimicked his rhythm as he watched intently, closer than I expected.

My face tried to project, *Unconcerned as Christian Ramirez.*

Marcus's face said, *Crestfallen at your ineptitude.*

When Marcus said, "Faster," I sped up.

"Faster."

"*Faster,*" he said a third time, followed by: "FASTER!!"

Mission: relaxation.

Status: tense.

"Stop, stop, stop."

I'd been at it maybe ten seconds. Sweat beaded my temples.

"It's about your ability to fire and relax, fire and relax," Marcus said, looking me in the eye. I had heard these words a hundred times and could recite them. Marcus explained that you start slow, fixated on the sound of your dorsiflexed sole popping the plywood. Boom, *boom,* boom. Left, *right,* left. Every triplet begins and ends with a foot dead center in the box. The ideal bounce springs from a stable home base, feet under knees, knees under hips, hips under torso. Then the foot that pops out wide smacks the angled wedge with audible force. (This, Marcus says, is one of the only P3 exercises that benefits from turning down the music.)

Boom, *boom*, boom.

Right, *left*, right.

Start slow, sit in the groove, and then crank up the speed—exponential acceleration. "Don't be afraid to fail," Marcus said. "The goal is to build it, build it, build it. Then you lose it and you're done." Ten seconds is enough.

Marcus demonstrated again. He was barefoot and sniffing sixty. His feet began pleasantly bouncy—boom, *boom*, boom—and quickly reached a pace that made the word *obliteration* pop into my head. He finished and smiled like James Bond with a fresh martini.

In the first few seconds of my next effort, he said, twice, "You gotta relax more."

*Why wouldn't I be relaxed?*

"Fire, relax," Marcus said. "Fire, relax." I took this to mean I should focus on both.

By the third attempt, I had forgotten Christian Ramirez, all strategy and plans, and instead just listened to the drum solo of feet on wood.

"Okay, that's good!" Marcus said as the whole thing blew up. "Good good good. Better."

Thinking he was making small talk, NBA prospect Anthony Black told Marcus one day that he had never been in the ocean.

"Anthony," Marcus said, "you should go in the ocean." It's about one minute from P3.

Could Marcus guarantee, Anthony wanted to know, that nothing would bite him?

"You're not going to get bitten by anything," said Marcus.

"Can you *100 percent* guarantee that?" asked Anthony.

"Ninety-nine point nine percent."

"Oh, hell no."

These anxieties, Marcus says, are "never just about one thing." A visit to the ocean would leave Anthony feeling newly empowered over the monsters of his imagination. Once Anthony believes he has the right tool for that job, Marcus says, he might believe he has the tools for other challenges, like moving his hips through a bigger range of motion, or his new job guarding the NBA's fastest players.

Marcus perked up when he heard WNBA top-overall draft pick Aliyah Boston say she wasn't the kind of person who went in cryotherapy chambers. Why go into her rookie year feeling inhibited? Doctors have a way of threading the needle between orders and advice. Marcus said she should start with just a minute, which she did. If the cryo chamber scares you, then going in it "doesn't just help you go in the cold," Marcus says.

Aliyah spent a lot of her time in Santa Barbara with another WNBA rookie, Haley Jones of Stanford. "Haley!" Jon yelled across P3 one day. "You look terrified!"

"A little!" she acknowledged.

Haley stood atop an eighteen-inch plyometric box, still as a chess piece, in the middle of P3's swirling activity. Her knack for scoring over and around bigger opponents delivered national championships, All-American honors, and a place as the sixth-overall pick in the WNBA draft.

With the words HUMAN > ATHLETE across the front of her T-shirt, and with arguably basketball's most recognizable ponytail, Haley was learning to keep a heel raised on her left standing leg on the box, while lowering her right foot, with control, to touch the ground, before raising back up to the starting position. All the while, she held a twenty-five-pound plate. The first time she tried it, her eyebrows strained together while she froze in thought.

It's a move that would challenge anyone with hip and ankle stability issues—but also, anyone anxious about testing hips or ankles. The idea that you need oxygen comes a lot earlier than the need for oxygen. Jon fetched a lower box. Haley would keep working.

Back when Alex led us through a breath hold, incoming rookies like Luka Dončić, Deandre Ayton, and Robert Williams III were about to visit in preparation for the 2018 draft and would practice breath holds. Results had been all over the place with the prior year's class. "You could tell a lot of them," Marcus said, "had never been in a calm place." To Marcus, that's proof that this breath exercise "is probably something they really need."

Movement is often the key to quieting our antsier feelings. Tim

Noakes's research found that our brains' central governors confine us—until you use high-intensity training, like sprints, to show the brain you can handle more. Trauma therapist and author Bessel van der Kolk explains in *The Body Keeps the Score* that bodies wracked by PTSD struggle with talk therapy, but heal convincingly with yoga or dance.

Rather than just needing surgery or medication, pain expert and psychologist Rachel Zoffness says many of her pain patients need to convince their brains that they can safely move. She has a practice she calls "pacing for pain," which she likens to training for a marathon. She asks patients to name a most beloved activity. You want to play soccer, but you can't walk today? Step one is to stand for three minutes in the kitchen. And on through the months, day by day, convincing your brain it's okay to do a little more.

When pain took over his body with a torn ACL, Marcus withdrew to his bedroom. Retreat is a common impulse and, Zoffness says, often ideal therapy for acute injuries. But chronic issues are different. If you stay home, if you stay in bed, if you don't see sunshine and friends, Zoffness says, you are not going to get better. "Exercise and movement," Zoffness writes in *The Pain Management Workbook*, "rewire your pain system, lowering the pain alarm."

Zoffness says pain is always caused by "biopsychosocial" factors. Medical science tends to look for issues in the tissues, but she says that it's not so simple. "It's never one or the other, it's always everything working together: biological, emotional, cognitive, behavioral, social, and environmental factors create the thing we call pain."

A host of factors can keep you from moving freely. The impulse box helps great athletes grow accustomed to relaxing through intense pace. "Nobody is a great mover," Marcus says, "that can't relax. You can get really strong, but if you don't relax, you don't have supple movements, you're not able to transfer between segments, you're always working against yourself."

All P3 staffers teach the impulse box, but perhaps not all as enthusiastically as Marcus does. They are also in broad agreement not to get a ride to the airport with Marcus—no one can relax through that. (Emotionally, they are on Team Nadine.) But life is an impulse box; so, at the end of one Santa Barbara visit, my son, Duncan, and I decided to test our relaxation skills.

Marcus's airport plan was that we would meet at Marcus and Nadine's around 10:30 a.m. for a 1 p.m. flight. It's about a twenty-minute drive. But Marcus had a wrinkle: when we reached the bottom of the Elliotts' road, instead of turning right, toward the airport, we'd turn left for a send-off swim in the ocean. "It's nice," Marcus said, "to get on the plane with a little sand on your skin."

At ten thirty, our bags were on the gravel under the live oaks. Nadine appeared a bit later, ascertained the situation, encouraged us to be very clear with Marcus, then left for a doctor's appointment.

The urge to ponder the security line at Santa Barbara International hovered like the urge to exhale in a breathing drill. I wondered if I should go to the bathroom again.

Marcus texted. Right after he dropped us off, he'd be lunching with a candidate for an important job at P3. He needed a few minutes to prepare.

We chatted with Keean. I took off my watch and put it in my bag.

Marcus emerged barefoot, asking if we needed food. Then he ran back to the house for towels, reemerged, and stood chatting on the gravel.

At last we piled into the Audi and swooped down from the hills to Yo-Yo Ma playing Bach. The weather was a postcard.

It was almost eleven on the dashboard clock, two hours to take-off, when Marcus mentioned he could use gas. Cool. We pulled into a station.

Marcus asked, "Have you ever driven away with the hose hanging out of your tank?"

I have not. Of course I have not.

"Have *you*?" I asked.

"I did it," Marcus answered, "a couple of times recently." Here's a clue as to why Nadine isn't at 100 percent on the Marcus-confidence Richter scale.

The most recent time, Marcus apologized to the attendant and got out his credit card. The guy wanted $1,500.

Marcus relaxed, inspected the damaged hose, and used his big Tony

Stark brain to see that it was engineered to snap back into place like Lego. He left his contact information and said to reach out with whatever documentation they had, so that Marcus could cover the cost of fixing the whole thing right.

Marcus never heard a thing. It cost him nothing. Somehow, the story of driving away with the hose makes Marcus look smart.

We parked in front of the sun-drenched Shoreline Beach Cafe across from Leadbetter Beach. Opening the tailgate for his wetsuit, Marcus stopped to chat with a jogger, getting a full download on that guy's triathlon training. Then Marcus mused about the many athletes in town on restrictive diets, which Marcus found a little off base. "We have to enjoy food!" At the café, a waitress delivered a massive pile of nachos.

Duncan picked up the thread and delved into findings about insulin resistance. Rapt, Marcus stopped wrestling his wetsuit for ten seconds, then a hundred, as Duncan unspooled some science. Marcus adores sober, evidence-based brains, and he uttered a most sincere *yeah* before resuming his neoprene battle. Eventually, we buttoned up the car, then Marcus squatted low to stash the car keys just so in a bush.

At last, we wended between two palms and squished along the soft sand. I had told Duncan we'd merely horse around in the waves, so he was wearing floppy shorts with flamingos all over them. Only now did I realize Marcus wanted to swim out along the buoys. Several days earlier, Marcus and Duncan had swum together for the first time and Marcus had ended up clinging to a buoy, catching his breath. He's cool about a lot of things, but not second place. When I beat Marcus in backgammon, he immediately texted to arrange another date, and another. Eventually, he proved to be superior and the urgency dissipated.

Duncan explained to Marcus that the cap, goggles, and jammers he'd need for a real swim were deep in his bag in the trunk of the Audi. Maybe bodysurfing would be fun?

Marcus looked at Duncan like he was speaking Italian. They ran back to the car, got the keys out of the bush, and tossed backpacks around. I waited on the sand as a fully naked one-year-old ambled in slow, sunny circles around his mom, on a blanket, on the beach. Camp counselors set up a volleyball net for campers in matching T-shirts. I made a tidy pile on the sand. Flip-flops on the bottom, then dry shorts, T-shirt, and towel.

Our flight took off in ninety minutes.

They returned at a run, all goggles, confidence, and Duncan's neon-pink swim cap. Before they plunged in, Marcus challenged me—a novice open-water swimmer—to work on pushing my chest low to be aerodynamic in the water and to swim to the third buoy.

When Marcus holds his breath, he doesn't picture a candle. He says the idea of floating in water is super-relaxing. *Actually* floating in the jolting cold of the Pacific, pelicans swooping low overhead, worked its own transformation on my antsy mood. Submerged, I quickly pivoted. We'd miss the flight or not. Let the day bring what it would. I dutifully touched the third buoy, swam back, and watched Duncan and Marcus lope in through the surf and hit the sand, grinning. The ocean tonic lingered through an outdoor shower, changing clothes, fetching the keys, and watching Marcus take off his wetsuit while discoursing on nuances of the kidney.

Before long, Duncan and I were buckled into row 19 of a United flight, a little sand clinging to our shins, high on the lingering Pacific skin-tingle. I texted Marcus that we had made it with eighteen minutes to spare. He apologized that he had been so conservative.

Somehow, I arrived to this flight more relaxed than any other. We think of the work of being an athlete as all grit. Marcus suggests it's also about smooth grace. "It's better," he says, "if your thinking part of your brain doesn't even get engaged."

# 17

## RATTLERS

*Poetry is nobody's business except the*
*poet's, and everybody else can fuck off.*

—Philip Larkin

MARCUS'S INNOCENT WORLDVIEW grew from the six hundred acres
of his childhood Eden. Marcus's parents turned him loose in the wilderness for hours or days at a time, supervised by a husky named Casey.
On night hikes, Casey would have the coyotes skedaddling before Marcus knew they were near. "Casey could always see," they said. Marcus's
parents made him fearless with a message few kids hear: "Nothing will
hurt you."

But, like Eden, there was an exception.

"We had a couple of natural springs," says Marcus, "and so we always
had running water when everything got dry. So, there's lots of animals
around, including rodents, gophers, all these guys that snakes survive
on." The family took a moral stand against killing, but made seven or
eight exceptions a year for rattlesnakes.

One got Casey. "All the hair on his neck fell off. He was sick for a
week or so. Just lethargic," remembers Marcus. Casey carried an enor-

mous black scar that reminded Marcus that "even the dog who was my protector gets taken down by these guys."

One day his dad found an enormous, five-foot-plus rattlesnake. "A beast," Marcus recalls. Later, they'd count an unlucky thirteen rattles. "My dad killed it," Marcus says, "but he didn't cut off his head like we normally do. Because it was such a specimen." With a bit of salt, and care, you could make the skin supple and beautiful enough to mount on a bulletin board or a hat. In the name of art, Gaius compressed the snake's neck with a shovel like an eagle would with a talon. Then he left the corpse on newspaper in the kitchen for skinning.

First, they made a quick trip to the shed to sharpen the knife. When they returned, the newspaper sat still and vacant, bereft of the monster with the thirteen rattles. "*Dude.* I'm, like, eight years old," Marcus says. His most fearsome enemy—slithery, injured, and angry, now loose in the inner sanctum.

"You have a sense that they're gonna fly in the air at you," Marcus says, "which is always what I feared." Five minutes seemed like an hour, half an hour felt like a lifetime. After forty-five minutes of looking, they glimpsed artful diamond-pattern skin at the other end of the house, under the couch in the dusk of the den.

They opened doors and began shooing, but the snake, Marcus says, "grabbed onto a leg of the sofa with his tail, and he's just like, 'Fuck you. This is to the death.'" Then the rattle began.

Gaius used the shovel to unwind the snake from the couch leg. "And he climbed up," Marcus says. "He inched his way up this corner of the wall. And he's bigger than me, as an eight-year-old kid. It was terrifying. I remember the slither up the wall like it was yesterday."

As much as Marcus studiously avoids inflaming problems like lost wallets or missed flights with the gasoline of anxiety, he'll bring his bike to a full stop rather than ride over a days-old rattlesnake flattened into the asphalt. His whole family knows a rubber snake will get him to shriek. "It's not well founded," Marcus says. "They just send shivers up my spine.

"Then my dad killed him. Cut his head off. Definitively."

That's the theme: except for the rattlers, Marcus's life is idyllic. Inside P3, the mood is lovely, a crucible of trust. In the wilds of professional sports outside the door, however, you have to beat the bushes and watch your step.

Low-grade corruption befouls much of the research into human bodies. "Most of the funding sources," Marcus says, "have some skin in the game somehow." Marcus's core conviction is that our bodies are brilliant. *We have the tools we need.* But there's no money in that. Research money often comes from people selling the idea you'd be better with surgery, pharmaceuticals, or gizmos. When Marcus researched race-day nutrition in South Africa, for example, it was funded by the Potato Board. Now that Americans bet more than $100 billion a year on sports, and venture capitalists pour about $200 million a quarter into start-ups aspiring to make waves in the growing betting industry, Marcus says he's approached all the time by deep-pocketed people with ideas about how they might use the data on P3's servers. Predicting injuries could be worth hundreds of millions or more in certain hands, as could knowing which superstar has scary ground contacts, whose hip range of motion has declined, or which defenders can move laterally with Anthony Edwards.

The Defense Advanced Research Projects Agency has been funding biomechanics research since 2001, for intelligence reasons. Our movement habits are as unique as fingerprints—and, unlike fingerprints, nearly impossible to keep private. There are hundreds of academic articles about how a dataset of movement patterns, like P3's, could identify someone in a disguise from the way they walk past a security camera in an airport.

Decades earlier, Harold Edgerton captured movement in exquisite detail. An electrical engineer, Edgerton ended up with the nickname "Papa Flash" for advancing high-speed photography in his MIT lab. Perhaps you've seen his stunning images of a milk drop striking a countertop and rising into a crown, a tennis ball at the instant one side is flattened by the strings of a racket, or a hummingbird frozen in midair. Edgerton captured a passing bullet so clearly, a critic accused him of hanging it by a thread. Like P3 and Muybridge, Edgerton decoded movement. "Anything that moves," Edgerton said, "is my game. It's always beautiful."

The military agreed. The night before D-Day, Allied planes flew over Normandy, fitted with Edgerton cameras to see Nazis like Casey the dog saw coyotes. After the war, the MIT professor founded a defense contractor, EG&G, which developed nuclear weapons triggers and other technologies they tend not to mention in his coffee-table books. EG&G operated a secret airport terminal in Las Vegas that was used by unmarked jets that turned off their transponders after takeoff. EG&G reportedly managed Area 51, and made sensors and electronics to let clients see warfare better. Under a new name, EG&G is now owned by the secretive Carlyle Group.

EG&G is a far cry from a biomechanics lab in a California beach town, but a longtime P3 client mused that P3 data "could be used for terrible purposes," and started talking about how a totalitarian state might funnel children into this or that sport, or the military. Many of P3's competitors—such as Fusionetics, north of Atlanta, and Sparta Science, in Palo Alto—count the military as core clients. A marketing video for Sparta leans heavily on retired general Spider Marks, who was the senior military intelligence officer during the LA riots and the Iraq invasion.

Marcus wants to keep P3 out of everything like that. "I feel like what we've done is built this academic research project that lives deep in pro sports." He calls it a "a perfect, unbiased environment." If new data upends a prior finding, they discard the disproven idea. P3's clients are athletes, so their focus is on "better information, stronger signals, and more precise outcomes to serve athletes. And how's the whole project funded? It's not by the Potato Board. It's not by the NFL charities. It's just by the athletes.

"These young talented athletes trust us with something so precious to them, and we take amazingly good care of it," Marcus says. He is disdainful of the politics and chatter that are many people's favorite part of sports. At one point, I brought up some NBA political drama that might play in P3's favor. Marcus cut me off, saying he had no interest in the "fickle politics and power swings of the NBA. It's one reason that I love our path; decoding the secrets of human movement transcends it all."

Marcus says he got an oddball business notion from his parents: either you're an artist doing pure work or you're slimy and greasing palms. P3 has had dozens of potential investors and partners through the years—Mubadala, Amazon, Microsoft, Chicago's sports-obsessed Ryan family, some early Facebook employees. He has often been close to bringing in someone more business-savvy to run things. The deals tend not to come together.

"Marcus," says Nadine, "is not a businessman." What CEO brags he knows nothing about his business's performance? "I don't want to know any of that stuff," Marcus says. "I have no idea what our books are, unless it's not working."

For a while, *no one* was looking at the books. When P3 first opened in the Funk Zone, Marcus felt weird working out in front of clients, so he'd slip away to Gold's Gym downtown. He'd park in a public lot, then cut through the Canary Hotel. One of the valets recognized Marcus. That valet started at P3 in 2010, and wasn't on the job long before he identified $75,000 in unpaid invoices. Adam Hewitt is now P3's general manager and the only person regularly thinking about marketing, business, accounts, taxes, or money.

Almost all of Marcus's closest friends are fine artists more focused on making beautiful things than money. "Acts of service and gifting are our love language," posts artist Nelson Parrish. "If this art career goes shitsandwich sideways, we are signing up to be Santa Claus."

"That is 100 percent how it feels," Marcus says. "It never feels like we're making something to sell or to market. It feels like we were making something that needs to be as true to our craft as possible."

At the Atlanta location, P3 lead biomechanist Leah Borkan has the job of explaining assessment findings to athletes, which is an exercise in translating SIMI data into plain English. Leah says things Marcus explains are "transposed onto that person and absorbed in such a wonderful way."

I ask how he does that. Leah has degrees in cellular biology and neuroscience, and a no-nonsense, "just the facts, ma'am" approach to problem solving. She doesn't tend to sound awed. But to this, Leah replies, "It is literally otherworldly to me." She points out that he's calm and never yells. But ultimately, she hangs his success on something like purity. "What he's built, wholeheartedly, he's doing it for that population, for the athletes . . . he knows he's on their side and they can kind of feel that."

But not everyone has the soul of an artist. In 2016, Matt Osborn stole a copy of *ESPN The Magazine* from an airplane because it had an article about Kawhi Leonard working on his hips in Santa Barbara. Osborn remembers thinking that the piece was intentionally vague, "like they didn't want to go into it too much. Maybe it was confidential?" Still, he smelled something special. "There's something in the weeds here," Osborn remembers thinking. "There's kind of a 'game recognizes game' among engineers.'"

Osborn had been a systems engineer for the Department of Defense, where he used 3D printing and outside-the-box thinking to innovate fixes for maintenance issues on nuclear submarines and attack helicopters. Then Osborn got a job with NASCAR's Joe Gibbs Racing, where other teams changed wheels, Osborn says, "with impact guns off the shelf at Discount Tire." Osborn's team designed a tool that could simultaneously remove five lug nuts, Osborn says, "faster than the human eye can perceive." Joe Gibbs Racing set all-time pit stop records; Osborn got a promotion and kept hunting down little technological advantages.

Then NASCAR banned Osborn's special tools, mandating that every team use the same impact wrench. "They removed this advantage," says Osborn, "and then that one. We had poured so much effort into it."

Osborn stewed, and nearly quit. Then he realized that they could take away a pit crew tool, but they wouldn't take away the pit crew. He had stolen a magazine off a jet that described people who could engineer human bodies better. This idea drove him, Osborn says, "right into Marcus's world."

P3's business model is based on the idea that keeping elite athletes active is worth a lot. The NBA workforce is a few hundred players whose combined salaries surpass $4 billion. In one study, NBA players were found to endure an average of 741 injuries a year that necessitated either seeing a doctor or missing a game. At any given moment, teams are paying millions to incapacitated workers. The injury crisis gets especially dire to teams' bottom lines when the players in street clothes come from the tiny set of superstars famous enough to influence national TV ratings. At almost every moment of the last few decades, at least one of the league's few true superstars has been out injured, while players like Zion Williamson represent the best hope to be the next LeBron, but haven't been healthy enough to achieve global renown.

Beneath the superstar layer, however, much of the sports industry assumes that if one athlete can't do the job, another will. "Next man up" means people are quickly expendable—both in the NBA and in the NASCAR pit crew.

"If your model is 'Kawhi is worth a hundred million to the Spurs, and we're keeping him healthy,' I have no interest in that," Osborn remembers telling Adam. "Our athletes aren't worth a hundred million dollars." Osborn needed to find people who were specifically good at changing tires and carrying gas. "I want to carbon-copy the few elite guys we have. I want you to tell me, from a systems mechanics perspective, why they succeed."

Osborn remembers Adam sounding "pretty lukewarm." But Osborn pressed. The NASCAR circuit would bring them to California—what if he came by with five pit crew guys?

Adam took it to Marcus. "I was like, 'NASCAR?'" Marcus says. "Don't waste my time, man. This left-hand turn sport is not our thing. We're trying to fine-tune these world-class machines. Not train auto mechanics."

Somehow, Adam convinced Marcus to join Osborn and crew on the pleather couches. Osborn explained that racing teams employ dozens of engineers and mechanics. Traditionally, those employees hold a footrace, and the fastest guys become the pit crew. Five days a week, you're an engineer or mechanic. On Sunday, you dead-sprint with 110 pounds of gasoline.

Joe Gibbs Racing had set pit stop records by hiring real athletes instead. But now they wanted to do better. Marcus perked up when Osborn told him his team had broken every pit stop down into 140 distinct movements. A good jack man, Osborn said, can have the jack in place on the peg half a second after the car stops. A tire changer can have the wrench on the nut one-tenth of a second after the car reaches a full stop, which means beginning the placement while the car's still skidding into place. It takes five-tenths of a second for what they call "the exchange," as one tire changer pulls the spent tire out of the wheel well and a teammate slides the next one in.

Can P3, Osborn wanted to know, identify people who can do *that*?

They spent most of a week working it out. "We started modeling the physical demands of these different positions in the pit," says Marcus. Osborn loved what he was learning. Eric's patter was all adduction, hip decel, and rotational athlete. "I remember looking at Eric," says Osborn, "and thinking, *I really don't understand*." But he learned.

P3 loved having a business partner who made decisions on a dime and spoke the language of data. And they loved clean wins. NASCAR's a time-sequence sport, like running or swimming. A pit stop is on the clock. It's a cinch to see results.

Over the years to follow, Marcus contorted his schedule to trail run or mountain bike with Osborn. Osborn and his family spent Christmas in the Alpine town where Nadine grew up. "We also started swapping books back and forth," explains Osborn. A shared passion was the work of mathematician Steven Strogatz, who writes about "one of the best ideas anyone has ever had," Osborn says, which is to use numbers not just to count sheep or determine property lines, but to explain the more complex aspects of our world, like circular shapes and infinity. Osborn and Marcus both advocate evidence-based decision-making.

In this, NASCAR had a huge advantage over football, baseball, or basketball: no priors, no entrenched thinking, no old-timers rolling their eyes. "It's a pure environment. Like Olympic sports, time series, they measure everything, which will tell you if what you did worked or didn't work," Marcus says. "That's so clouded in team sports." In NASCAR, the systems engineers pursued whatever worked.

For the tire-changer job, they'd recruit college baseball players who just missed the professionals. The tire changer carries a heavy thing while pivoting at the torso. "He's a rotational athlete," says Marcus. A tire carrier has to pick up a heavy item and run. "The tire carrier is a max force athlete, the Zion of the pit," says Marcus. They hired defensive backs and running backs for that job. The best jack men in the world, they found, are big enough to have powerful leverage on the long jack handle, but small enough to move from one side of the car to the other at Olympic sprinting speed. After some trial and error, it emerged that the best performers were essentially all the same height: six foot two.

Joe Gibbs Racing would identify job candidates and give them learning aptitude tests (because almost no one grew up playing the sport of NASCAR pit crew, every recruit would need to learn their movement from scratch). They sent all the remaining candidates to Santa Barbara, and hired whoever the P3 assessment said had the movement profile of a champion.

The partnership upended everything. "Instead of four years to reach a professional level, now it takes three to six months," says Osborn. "Every year, we have a new Tom Brady."

In 2022, after a few months on the job, former arena football running back Jake Holmes made his NASCAR debut as a tire carrier in the sport's premier race, the Daytona 500. In that first outing, Joe Gibbs Racing credited him with the fastest tire-carrier performance ever. Around that time, an organization called Asphalt Analytics emerged to track pit crew stats; Holmes was rated best in the sport.

By the end of the 2022 regular season, according to Pit Stop Stats, there had been thirteen pit stops in NASCAR history faster than nine seconds. Ten were by Joe Gibbs Racing. The same year, Joe Gibbs Racing set a record for the fastest pit stop of all time, with an unthinkable time of 8.6 seconds.

Joe Gibbs Racing would seem to have a bigger scouting challenge than any other sport: the athletes they recruit have not played the sport. But Matt suggests that "our hit ratio might be higher" than the NFL or NBA.

How could that be? Marcus thinks the answer is that it's the only sport that instantly made biomechanical data central to decision-making. "This shit is so good," Marcus says, "everything else is just dead weight."

There was one part of P3's relationship with Joe Gibbs Racing, though, that wasn't totally pure. Osborn says, "We use the data for nothing Marcus intended." Matt remembers noting that P3 was really focused on injury prevention. Osborn mentions "quality over quantity," which I can only interpret as quality over quantity *of athlete*.

"We're the opposite," Osborn says. "Our athletes are somewhat expendable." Osborn's deep pragmatism can feel mechanical, even ruthless. Joe Gibbs barely used P3 training, and didn't put a big emphasis on injury prevention. Osborn says, "We're kind of the bottom-feeders of the pro sports world." For athletes on the fringes of sports, losing a solid Joe Gibbs job can be a body blow. "We were starting to make people feel insecure. The turnover was so high, the pace of improvement so fast. The Tom Brady of the sport from last year is wondering how to save his job."

Osborn says Eric told him that the Joe Gibbs job candidates appeared to be manipulating the assessment. If you're taking motion-capture and force-plate measurements to refine performance or prevent injury, it makes no sense to be devious. If you're screening for a six-figure job, though, you might try to hack the test.

First, the P3/Joe Gibbs collaboration made athletes disposable. Then it made P3 disposable. Osborn used his brilliant brain and his fast, efficient decision-making to turn Joe Gibbs Racing into a P3 competitor. Osborn called me from the site where he was overseeing the construction of a $4 million facility with force plates in the floor and motion capture all over the ceiling. Osborn calls it "the most advanced biomechanics lab in North America." There's a 60-by-120-foot area with 135 force plates in the ground where people can play sports. "Instead of days to break down data from one person, ten guys play basketball on a court, and you can see all the same data from every limb they move, live during actual movements during the event," Osborn says, adding that they're lining up partners like the 76ers, the Mercedes Formula One team, and the Bundesliga. Osborn tried to hire Eric away from P3, failed, and then made Eric's deputy, Trent Reeves, the first director of biomechanics at the Joe Gibbs Human Performance Institute. When I saw Matt speak at a conference in early 2024, the front row was loaded with the same Charlotte Hornets employees I knew Marcus had been courting as P3 clients.

Marcus meets often with professional sports organizations who get confused into thinking that P3 is a motion-capture business. Late one morning, Marcus talked for most of a bike ride about a meeting he had just had with an NBA team. They described the gizmos they had purchased and the data they would collect. But in essence, they still lacked all three of P3's crown jewels: a mature database, specialists to extract bankable findings of human movement, and the doctor steeped in injury prevention to hire and run the training staff. "How will you know," Marcus says he asked, "what to tell an athlete to stop doing, so he won't get hurt?"

# 18

SLOW

*The time lag between research and practical
implementation in general healthcare
settings can be as long as 17 years.*

—Arundale et al., *Journal of
Orthopaedic Research*

THE JOE GIBBS HUMAN PERFORMANCE INSTITUTE wasn't Marcus's only new competitor. In 2023 dozens of new biomechanics shops appeared— many of which sought to become the NBA's league-wide biomechanics partner. Nine years into Marcus and P3 assessing players at the combine, the league had taken Marcus's advice and won, in collective bargaining with the players, the right to biomechanically assess every NBA player four times a year. Now the league would pay someone to do that work.

As the near year-long process wound down to a handful of finalists, Eric saw winning that contract as critical to P3's business stability, and Marcus felt dispirited.

Marcus has stories about how much time he spent explaining his life's work to Breakaway Data, for example, another finalist. Breakaway teamed with P3 for a while, then raised money from a private equity firm with deep connections in sports—Elysian Park Ventures ("We invest where sports intersect with life"). "Next thing we hear," Marcus says, "they're basically pitching to do a lot of what we do. And we've shown them everything we do."

The league brought in more and more outside experts and consultants as the process unfolded, demanding new proposals, presentations, and a command performance of P3's SIMI motion-capture system—as conducted by people with no expertise in using it.

Marcus says that data collection was the easy part. P3, he said, would work with any system the NBA chose. The vastly trickier business was making athletes more robust. P3 can see who gets hurt, and then look back through years of assessment to see how they move. P3's models can identify six out of ten athletes who'll suffer catastrophic knee injuries. Another study demonstrates they can do the same for lateral ankle sprains. But the point of prediction is creating an opportunity for intervention. The end goal is prevention, which is far trickier to measure.

In the spring of 2024, the NBA picked an amalgam of vendors to execute the new program, none of which were P3 and one of which was Breakaway Data. Marcus found the whole process dispiriting and wrongheaded, bad for athletes and bad for P3. But then Marcus went for a solo night hike up a creek to steady his emotions, and resolved to send Eric to attend the draft combine—where, for the first time in a decade, P3 would not be assessing players—to see if he could be helpful. In Chicago, Eric learned something shocking: due to a clerical oversight, the NBA had never read the core of P3's submission.

Marcus took it as "probably a good thing," because it cracked the door to one more round of conversations. "We're in discussions right now on what that would look like," he says, "and we're going to play some role. I can't imagine us not playing a role, even if the league didn't pay us. We want to help them, you know, try to get it better, at least."

One worry Marcus has is that if the league initiates entirely new kinds of assessments, it'll take three to five years for the database to start spitting out meaningful information about injury prevention.

Over the months of this process grinding along, an NBA player tore his ACL. As is P3's custom, they dove into their data to see if that player had known risk factors. He did not—but he was fresh off an ankle injury. P3 learned that, as he returned, the player had severely reduced ankle mobility. This presented a plausible mechanism of injury: the ankle stopped absorbing much force; the knee bore the brunt.

And that's where P3 had insight the league won't get anytime soon: all-time, there have been six ACL tears from NBA players who did not first demonstrate ACL risk factors in their P3 assessment. Do you know how many of those torn ACLs were in players fresh off an ankle injury? *All six.* Do ankle sprains elevate ACL injury risk?

Even though his own ACL tear came from a large teammate falling on his leg, Marcus sees that injury in a new light now that he's marinating in movement data. He had broken his right ankle about a year before tearing his left ACL. After trying to play on it for weeks while it was misdiagnosed as a sprain, he healed slowly. When he finally came back, he says he did so with an ankle that barely moved. When he tore his ACL, he couldn't shake the idea that his ankle injury was somehow related, "and what I know now is that almost certainly it was. The mechanics of ankle injury . . . almost always set your knee up for bigger ACL risk. As you max out your ankle mobility in flexion and dorsiflexion, you can't get this additional six or eight or ten degrees." Marcus says it routinely affects the opposite knee: "Big limitations of ankle mobility on one side almost always end up with excess valgus and sometimes rotation across the long bones also, which is really dangerous."

Making these kinds of connections requires more than a dataset. "People are collecting an enormous amount of data without knowing what to do with it," says Charles Kenyon, DO, the former P3 intern who is now a sports medicine doctor. Marcus has a story from Phoenix to illustrate this point. In 2019, the Suns opened what purported to be the most advanced training facility in the world.

"Tucked away in the nooks and crannies of the courts and workout areas are 150 HD cameras, sensors, nodes and 3D force plates that can track the motion of players and the ball," says the press release. "Using Verizon 5G Ultra Wideband connectivity, the Suns are merging computer-aided motion analysis, player and ball tracking, and shot tracking—three usually isolated technologies—into one integrated system." The "revolutionary 5G system . . . could change the very nature of NBA training forever."

Then a funny thing happened: Two and a half years after opening the facility, the Suns changed hands, signed Kevin Durant, and called P3. P3 staffers flew to Phoenix and spent several days getting to know the facility and the staff. They called Marcus with news.

"They don't know how to turn it on!" Marcus reports with a laugh. It appeared the much-ballyhooed wired court had not been used, and had deposited no data on any servers.

The Suns' performance leaders have degrees in athletic training, sports conditioning, and rehabilitative sciences. They're used to stretching and squatting players. Without intricate knowledge of the human body, without a scientific mindset grinding on a mountain of data, you're lucky if your motion-capture system can progress past the floating pelvis stage.

It's strange to see the sports world operate largely without the insight of well-analyzed movement data. On a Wednesday in May 2022, the brain trust of the basketball world walked into the gym at UCSB at nine thirty in the morning to see the players who'd been training at P3 for seven weeks. In attendance were William Wesley and Tom Thibodeau from the New York Knicks, David Griffin from the New Orleans Pelicans, the Warriors' Bob Myers, the Los Angeles Clippers' Jerry West and Lawrence Frank, the Dallas Mavericks' Nico Harrison, the Nets' Sean Marks, and two hundred others like them.

A month earlier, I had asked P3's Adam which of the players they had in Santa Barbara I should know about. He pointed immediately to a player who, on that day, was considered a fringe first-rounder: Santa Clara's Jalen Williams. He had perhaps the longest arms ever measured on a guard, and tested in P3's rarest body type category: "specimen."

But as organized by agents for general managers, the Santa Barbara showcase didn't present Williams as a star. He played in some early three-on-three and then took a seat as bigger names took the court. Few power brokers grilled P3 staffers about prospects. One GM told me he was only there to "kiss the ring" of the powerful agents on hand. When the Oklahoma City Thunder picked Williams twelfth, it was seen as too high; a few years later, it's clear he was drafted far too low.

Chet Holmgren trained incredibly hard at P3 and measured well.

Marcus says the men who run the Orlando Magic and had the top-overall pick told him Chet was too skinny. Then Chet missed his whole rookie season after an injury and the Magic looked prophetic for passing.

But that injury, Marcus says, came from the freak accident of LeBron James stepping on Chet's heel in a charity game, resulting in a Lisfranc foot injury. "That had nothing to do with how skinny he is. But I guarantee the president and the GM of the Magic are saying, 'Yeah, we saw it. We saw it.' And now they're gonna be likely to make the same mistake over and over."

When he finally got on the court a year later, Chet had one of the best rookie seasons in NBA history, and finished second in Rookie of the Year voting to the arguably skinnier Victor Wembanyama.

The prospect of biomechanical assessments at the center of the sports world threatens trainers, executives, and also doctors. When I interviewed Kenyon in 2023, he said he had just been at a medical conference where people were still talking up a 2016 paper called "Why Screening Tests to Predict Injury Do Not Work—and Probably Never Will . . ."

"It's a pretty catchy title," says Kenyon. In logging long hours in Major League Baseball doing injury prevention, Kenyon met many people wedded to the old ways. "A lot of people point to that story," he says, "as evidence that screening doesn't work."

The *British Medical Journal* paper is by Roald Bahr, MD, one of the most decorated sports science researchers in the world. Prince Philip once presented Bahr with an award at Buckingham Palace; he's the chair of the medical department at Norway's National Olympic Training Center.

When I asked about Bahr's paper, Eric made sure I noticed the first letter in response, from Timothy E. Hewett of the Mayo Clinic, which contains the academic equivalent of a sick burn. "His data," writes Hewett, "are most likely not Gaussian in nature." Hewett notes that Bahr is on record as recommending screening for soccer players, for basketball players, for volleyball players, for people with poor landing mechanics, and for frontal knee kinematics. "In multiple publications, Dr. Bahr has extolled the virtues of screening," writes Hewett. "These statements raise concern regarding why Dr. Bahr has reversed course 180 degrees after his own studies validated screening."

Hewett concludes that screening "can clearly be used to identify risk subgroups."

But Eric's feelings were more nuanced. "Everyone acknowledges," he says, that we'll never get to a point where he can say, " 'You're going to blow out your right knee in seventeen and a half days.' " The hard work will be to see how precise those screenings can become.

"What he's doing, in some respects, is setting a very high bar," says Eric. "And I don't think that's a bad thing. Injuries are awful. And setting some pretty stringent thresholds to be able to allegedly catch these things before they occur seems like a relatively good practice," especially in the loosey-goosey world of sports advice. "A lot of the training space lives in the medically adjacent zone where there's not much regulation," says Eric. "There's a lot of *feel* involved. Having this be pretty systematic—it'd be nice to err on that side. You don't want to claim injury prediction or prevention when, in fact, you're kind of full of it."

Bahr points to a medical standard called the Wilson-Jungner criteria. They come from a 1968 World Health Organization paper that purported to address an important question: Given limited global resources, which health problems should be addressed with preventative screening?

The correct answer might seem to be: as many as possible. But it's taxing to ask doctors to test whole populations for anything. Over time, the Wilson-Jungner criteria have been refined, and now recommend screening only for problems that are: 1) serious, 2) detectable in an early stage, 3) benefit from early treatment, and 4) have a suitable test.

Those are stingy criteria. When Wilson and Jungner were writing, mental illness, breast cancer, and high blood pressure didn't make the cut. They expressed concerns about diseases with complex mechanisms like diabetes, tuberculosis, and Følling's disease.

Eric wonders, though, if biomechanical screening might be simpler. "We understand, generally, how a body is put together," Eric says, "how certain ligaments attach, where certain muscles attach. We can start to put together mechanisms of injury that even just mathematically, we can work out what mechanisms would put a strain on an MCL or an

ACL." Sometimes you can see how a foot endangers a knee. Marcus often says that P3 does "high school physics."

Eric says that, at his desk in the back of P3, they find themselves seeing "a plausible mechanism at play here that would increase the risk of right hip issues. We can probably start to knit that together with some data that we collect from a movement screen that will tell us whether or not a certain portion of the musculoskeletal system is going to be stressed. I don't know that the mechanism of pinpointing breast cancer onset follows the same rules."

With each passing year, the data seems to be more clearly defying Bahr. P3 might be on the bleeding edge, but it seems inevitable that conditions ranging from ACL tears to lower-back pain will soon meet the Wilson-Jungner criteria, if they don't already.

What would sports look like with screenings baked into the routine? While Bahr was writing, former P3 employee Jeremy Bettle was discovering the answer. Bettle left Santa Barbara in 2011 and went on an eleven-year tour of the major leagues, working for the Brooklyn Nets, Toronto Maple Leafs, Anaheim Ducks, and the New York City Football Club.

The experience left him with something approaching post-traumatic stress disorder because of the institutional resistance to evidence-based changes to athlete care. "The power dynamics," Bettle explains in his British accent, "are spectacularly skewed." Bettle says, in grave tones, that there are stories he won't tell—and notes that research and reason struggle in organizations run by an all-powerful few.

"I'm going to come in with my laptop and go up against the coach who's making $10 million a year, and there's only going to be one outcome if we start getting into a fight."

"It's not good for your laptop," I say.

"It's devastating for the laptop."

Bettle spells out a hypothetical: a player is returning from ACL surgery. Six months after the procedure, the surgeon tells the GM that the player is good to go, because the grafted tissues look healed on a scan.

This talk, between the highly paid surgeon and the highly paid general manager, happens, Bettle says, "with zero regard to the detraining level of that athlete. And so now it's on the performance director to go and tell the GM, 'Well, actually, no, he's not cleared for another six months.' And so you immediately create conflict."

Surgeons often give patients a rehab calendar: a certain number of weeks until non-weight-bearing exercise, then weight-bearing, then running. It all ignores how the player is moving. "Based on an arbitrary number of weeks, okay, good, you're cleared to run," Bettle says. "But before that day, you haven't been allowed to take them in the weight room. And so they go from nothing, essentially, to a thousand single-leg plyometrics, and then they get sore. And then we have to shut them down for a couple of days. And then we let them run again."

Bettle says the pattern is "get injured, massively underprepare, return to sport, get reinjured, massively detrain again, go back early to get reinjured." And the worst part is "people just sort of accepting that an athlete will never be the same after an injury. *It's because of how you're rehabbing*. They become injury-prone, because we don't give them the time to recover. And we don't give them the time to repair the mechanics that caused the injury in the first place."

What would be better? "An objective, milestone-based process," Bettle says. He implemented exactly this sort of system when he arrived at the Toronto Maple Leafs in 2015. The team's physical therapist, Ryan Morrison, dug into the literature around the groin injuries that bedevil hockey. They assembled a trove of data points. "And then, working from those," Bettle says, "we came up with four or five measures that we did every couple of weeks."

Screenings. Screenings all the time. "It was essentially strength in different hip ranges of motion, as well as force-plate and training load data. And so we could see if an individual athlete's, let's say, hip internal rotation started reducing on their left side over a two- to four-week period. We knew that that was increasing their susceptibility.

"And so the intervention would come with the physical therapist working at the joint level, and then the strength coach working at the muscular level to actually retrain that range of motion and make sure that we had strength through that range. And then it would go on to the development coaches, where we would actually integrate that into some on-ice work that we wanted the athlete to do. Maybe it was skating technique that was causing this, maybe it was the side of the ice they played on." If your groin didn't like playing on the left, they'd switch you to the right for a while.

"You get to a point where you can start stratifying into groups and why they're at risk, and then you get an intelligent system of load management, knowing that everyone with weak adductors, everyone in the bottom 30 percent, would be our group who are susceptible to groin pain and injury."

"Ooh," I tease, "that weak-groin club must have been a cool club to be in."

"That was the group everyone wanted to be in," Bettle confirms.

But the reality is no joke. "After six months of the system being in place, we basically didn't have another injury, a groin or anything else. Nobody missed for a noncontact injury in the rest of my time there. It was unbelievable. Proud of the group effort, you know, because it took all of us.

"It shows," Bettle says, "that injuries don't have to be a part of sport."

# 19

## GREAT HORNED OWL

*You have tried your powers too little.*

—Nikolai Rimsky-Korsakov

IN 1955, WHEN Paul Dudley White was sixty-nine, he wrote a cover story for the *New York Times*. In an accompanying photo, he's tromping through the woods in a suit, tie, and fedora. A medical journal says 50 million people read his story. "I myself," writes White, "have found that physical exercise such as a long walk or bicycle ride or wood-sawing or log-splitting, or an interesting book, or music, can be a helpful antidote to nervous fatigue and worry over patients. Incidentally, I have just returned much refreshed from an eight-mile walk over Newfane Hill in Southern Vermont, where I am now writing down these reflections."

In a 1957 follow-up, White proposed regular exercise for every person. He told of a 103-year-old friend whose swollen legs had been cured by walking, then wrote that "it is time for us at all ages to be more than spectators of the sports of the day. We should expand our physical activity beyond that of getting into and out of automobiles and riding in elevators, buses, trains and planes."

When White lamented what he saw as insufficient funding for research to investigate the relationship between exercise and heart health, his words fell on powerful ears. In 1956, White's most famous patient, President Dwight Eisenhower, established the President's Council on Youth Fitness, which set the government on a path to recommending that regular Americans move more.

A year before White's death, the American Heart Association recommended a half hour or so of endurance exercise several days a week. The American College of Sports Medicine took a similar position in 1978, and at long last in 1995, the federal government began recommending regular exercise for Americans.

Today, it's hard to find any expert, gym teacher, or family doctor who doesn't sing from this hymnal. As of 2024, the US Department of Health and Human Service recommends that adults get two and a half to five hours a week of moderate-intensity activity, or half as much "vigorous-intensity aerobic physical activity."

Marcus follows the guidelines with constant bike rides, runs, and hikes. But he also notes that it's called "cardio" because it's good for your heart. For too long, exercise guidelines have ignored the rest of the body. At a picnic table among the potted succulents in the garden of a Santa Barbara brewery after a long bike ride, Marcus said it's time for those guidelines to go.

We already know they haven't motivated people to move as directed; the CDC says less than a quarter of Americans follow exercise guidelines. Perhaps because they don't address the root of the immobility crisis: people are more likely to move bodies that feel good.

"You can go get on your elliptical trainer, and you can go get on your stationary bike, and you can do some exercise and not be really out of shape," says Marcus. But, he argues, it's a rudimentary, and ultimately doomed, approach to solving the issues that most trouble athletes, like ground contacts and hip mobility and stability. Twenty minutes on the elliptical, Marcus says, "can have these systems really go to shit, because you're doing nothing: none of that complex navigating in space, ballistic movement. And frankly, when I think about things that civilians can have, *that'd be nice to have.*" To Marcus, a highly functional set of hips, and springy ground contacts, are the targets.

Those aren't standard goals. I sent Marcus a newspaper story with the headline "Whatever the problem, it's probably solved by walking."

"If you're talking about improving the health of America, I don't entirely disagree. Many things will get better with moving," Marcus texted back. "But it's a pretty dumb answer to a complex question. That approach will help some middle swath, but will also miss so many."

"Most people," Marcus says, "by the time they're forty, they're just brittle. There's no bounce in their body. There's no impulse to their bodies. There's no relaxation, spring." This, he says, is what we all should strive for. "You can definitely keep a lot of that around by training."

Now we're getting into the grand questions of Marcus's life. "How," he asks, "do you care for a body that's going to be a ballistic body? That can do amazing things? Not so you can do a little bit of aerobic exercise, to not develop an increased risk for cardiovascular disease, but so you can do crazy, exciting, engaging, amazing shit right now?"

Marcus suggests we take inspiration from animals that move a lot: they play. One way to get humans playing is team sports. Another is outdoor adventure. Marcus's weekends are improvised from those elements: screaming up and down mountains, jumping in and out of water—then racing off to arrive just in the nick of time for whatever basketball, volleyball, and especially soccer game one of his children is playing next.

On those sidelines, Marcus conducts something of a sports confessional. On a glowing green fake-grass field at UCSB, friends and acquaintances circle by with an endless report of whose collarbone is healing slowly, who dropped time in the triathlon, who didn't get a college scholarship because of some dumb coach, and who's off crutches. Marcus gets a lot of medical insight from P3's assessments, and another big dose from walking around.

So, Marcus noticed when he began hearing a new kind of story. Amid the sad tales of rotator cuffs and MCLs, he heard tales of parents, in middle age, flicking on some long-lost switch. And doing it in maybe the simplest, almost silliest way possible.

Pickleball.

A game named after a dog named after a condiment sprouted from the space between tennis and Ping-Pong. One couple, Marcus says, had never been athletic, then started pickleballing six days a week and talking to him about it on the seventh. Marcus sensed awakening. The message, he says, was "Look what I can do!"

"It's the same thing we feel when we're running down some trail and dodging over rocks or a riverbed," he says. "It's amazing for these people that haven't done anything with their body for twenty-five years to get out there and see how it can navigate in space." Elliott isn't prescribing pickleball for everyone. He has a tennis player's snobbery ("Pickleball's such a contrived, silly game"), but he likes that it's zipping electricity, endorphins, and oxygen around once-dormant limbs. To Marcus, it's obviously helpful that the game makes people cackle with delight.

Middle-aged runners and bikers, Marcus notices, can be a little more haggard. Every runner in the master's age group knows, likely to the second, how much speed they have lost. A twenty-five-minute Turkey Trot is wonderful, but a little less high-fivey if you used to run it in seventeen. The clock can be like glaring light in a photograph, highlighting wrinkles.

Amorphous pursuits age better. A pickleballer of almost any age can hit a memorable winner. On a hike, Marcus interrupted the conversation to point to a craggy ridge in the distance. He closed one eye like a sea captain and lined up his pointer with immense precision. "Jonas and I camped," he said, "*right* there." Marcus was proud. But also it's a dot in a wild swath the size of Massachusetts. The mountains offer infinite victories.

There's a spot way out on the edge of Death Valley that's sand dunes, mountains, hot springs, and dramatic skies. When Marcus, Jonas, and friends arrive, they dress however they must to survive the elements. (To Nadine's chagrin, Marcus often ends up in polarized sunglasses and a sarong.) Someone picks the next hole for insane Aerobie disc golf, across miles and canyons. You see that cactus a mile away, across that chasm? Par-22. And off they go, scrambling through wilderness, rock climbing up, leaping down, dodging scorpions and rattlesnakes. (Jonas has a story about an electrical storm that left him coated in luminous, plasma-like glowing balls of electricity, which sounds like St. Elmo's Fire.)

"How," Marcus asks, "is that not more popular?" Mostly, they laugh their asses off.

The church of human movement exercises a full range of rituals, honoring our evolution as unusual, but elite, movers of the animal kingdom. Once you value getting low, stretching long, quivering with anticipation, exploding into attack, and clinging to a tree branch, you start noticing the mind-blowing physical competence of other species. Squirrels, cats, and foxes are Jordanesque.

My back injury had made me less like a cat and more like a cow. Bovines can't walk down stairs, and only jump over the moon in myth because it's hilarious. They're mostly head down. Our word for curling in on oneself in fear is "cowed," and I have felt that. Mountain biking through sunrise in the early days of my back twinge, I rode my brakes and anxieties down a winding, rocky descent. By the time the path flattened out, I had been ditched by my riding companions.

But that's not to say I was alone. A hundred yards to my left, at a time of the day when people mostly pour coffee and brush teeth, a man dressed for work and a woman in a robe stood sunlit on their lawn, peering into the shadowy woods. She waved an arm; they had the tense vibe of citizens in a zombie movie.

I was mashing the pedals, glasses jostling, ears crowded with exhales and wind. Was she yelling at . . . *me*?

The path curved and swooped, the brush melted away, and soon they were close, obvious as cabaret stars in their spotlight, and definitely trying to get my attention.

Maybe fifty yards away, I put two feet on the ground and hollered, "WHAT?"

Both yelled; sounds muddled in urgent confusion. But just at the end, her last two words hung in the air: *YOUNG MALE*.

The base of my neck tingled.

*Ta-da!* Rising from a gulley between us, jogging directly at me, was the first wild black bear I had ever seen.

I recall every detail of that impeccable face. His almost-black coat tapered to lush mahogany around the snout, like a Rottweiler, with almost every hair smoothed to Pat Riley perfection. Was he smiling? I saw his tongue rest in the canyon between canines.

Black bears, say the experts, don't want trouble. Get big, loud, and proud, and they'll cede the ground.

Instead, I pedaled my freaking ass off. I booked.

The thinking is that if you run away, your prey-like movements might excite the predator part of the predator. I was so certain I was being dumb that I actually turned to holler, "I'm just going to get out of here."

I was thinking that the bike would tilt the calculus—and it might have, on a smooth gentle downhill, where I can move like an ewok on a stolen speeder. Alas, this trail turned uphill. A little at first, and then more. My heart and legs fought and fled, but the trees didn't whiz by like the forest of Endor. Then I reached the switchbacks.

Cutting a trail back and forth across a hillside is normally merciful to bikers. It's easier than climbing straight up at a steep angle. But that day, I felt like a passed hors d'oeuvre. I wove at a snail's pace left and right and left. A young male could walk right up the middle—if he missed me on the first pass, I'd circle back. My posture grew cowed, cramped, my shoulders hunched, my lungs tight.

At the top of the hill, my buddies leaned, alive, on their bikes. I blurted my bear news. Then the big fellow lumbered into view and crossed the road a little way down the hill. He had skipped the switchbacks and followed the stream instead of me.

We finished our ride glued together into a herd, hollering, "Hey bear!" at every corner. I had thought good biomechanics were about freeing your body of pain and weakness. But when the bear rolled through, I ceded my posture, lucidity, and breathing pattern to the threat, which strode off in a shiny cloak woven of my deference.

That's when it struck me that the freest movers tend to be at the top of the food chain. Ospreys, owls, falcons, and jaguars have a dreamy, giant range of motion in the joints, a maximus posture, and steely gaze. Deer and horses can run, but the cheetah's magical ground-pulling paws are poetry.

Owls might be the freest animals. They glide through the night on wings that baffle scientists with their perfect silence, and feast on snakes, cats, young foxes, and even bigger raptors. Like many professional athletes, they're not big on nesting. If they can't steal one from an eagle or an osprey, they'll just lay their eggs on a stump or the ground. The BBC has footage of a wolf pack slinking toward owlets on the open tundra. The adult owls zoom in silently, sinking talons into wolf haunches from behind. At first, the wolves leap at the owls in anger, and then, after a few more attacks, jog off in agony.

I think about the upright spine, the open chest, enormous eyes, and oozing fearlessness of an owl. And I think about my hunched shoulders, craned neck, and shallow, panting breaths as I fled the bear.

An Oregon nature writer has a story about scrambling close to a great horned owl nest in a fallen tree. He ignored the owls' warning clacks and ended up in the emergency room, getting his ear reattached.

They're named for the "horn," an extravagant *V* of feathery tufts jutting from their crowns, but they lord about in absurd oligarch fur coats. Marcus told me a story about driving home one dusk years ago, when the headlights caught a great horned standing on the road. He runs out of words to capture the feeling of the cream-and-black chest feathering, an haute couture tumble of maitake mushroom. Describing it later, Marcus squinched his eyes, puffed out his chest, drew his hands out like an opera singer, shook his head slowly in a hoo-boy manner, and delivered a sound like the purr of a dragon. It's a sound that expresses, he says, that this bird is "not something normal. It's something bigger, it's all turned up in a way that human language is not capable of expressing." He pauses, then adds, "I'm trying to say *that shit was good to go*." Here lies the highest physical standard.

Just as Marcus drank in the eighteen-inch owl, a truck crushed it. The thin-boned beauty took the brunt of a few thousand pounds of aluminum and carelessness. A hit-and-run. Marcus got in the road to take in the tragedy. He fetched an open-topped cardboard box from the car. This was before they had kids, when Nadine and Marcus lived in a rented casita above a rich person's garage high in the hills above town. Marcus put the owl in the box on the porch. Nadine called the bird rescue place. They both wondered if it would survive.

The next morning, though: big, open owl eyes. Nadine put the box on the passenger seat and talked to the owl. "He was not out of control," she remembers. "Just steady." The rescue people said he seemed to have a broken wing and didn't chat as much as Nadine wanted.

Maybe two weeks later, Nadine remembers hearing owl sounds; not a distant nighttime *hoo hoo*, but house-shaking, midday loud.

The casita had a patio, two chairs, and an unfuckingbelievable view. And on that day, on the black metal railing: a great horned owl.

"Can it be that it actually came back to our house?" wonders Nadine. She believes so. "He found us! He just sat there. Marcus came outside, and he just looked us in the eyes. Have you ever had a bird just sit still and stare you in the eye? He took off after thirty seconds or so."

Nadine called the rescue place, whose records said their owl had been released at the scene of the accident, just down the hill, a day or two before. "The guy," says Nadine, "said that would not surprise him one bit" if the same owl had returned.

It was a big excitement, a few months into physical therapy, when Jen said I could mount the elliptical. Up there, I felt like a one-winged owl in an open-topped box. At least I had a view. From that tall perch, it struck me that the room had P3's basic size, lighting, and floor plan. The equipment was familiar—pulley-based total gyms, resistance bands, medicine balls, stacks of towels. Every person had their own detailed medical file and customized exercise program. The staff had advanced degrees and supervised just a few people at a time, with abundant time to prescribe a custom workout, listen to how it was going, tweak, and tweak some more. Sometimes there were half-hour conversations about how a body works, an owner's manual of movement. There are facilities roughly like this all around the world.

The difference is, you don't get into this place by being an elite athlete, or even a mediocre one. The price of admission is to be *broken*. The instant you can start to move normally, you're disqualified by the insurance company. What a crazy gym! A seated woman with jet-black hair and a sari strained to bend an undeniably phallic tube of rubber. A tall man with a gray beard walked carefully up two steps, and then down two steps. Two women on tables clamshelled banded knees. PT is not P3.

This place had neither NBA players nor systemic movement data—only a radiologist's scan and the observation of expert eyes. But maybe the biggest difference was in mission: the physical therapy gym gets you back to work or school. P3 wants you to fly.

Over many weeks, I progressed to movements like walking with a light kettlebell held upside down above my head, and monster walks—big, lunging diagonal steps with a thick rubber band around my ankles. (It's a Stephen Curry–esque reminder of the conspiracy between hips and ankles. The resistance is at the floor, the work is in the hips.)

To keep from monster-walking into another patient, I'd often retreat to a sunlit side room that had the aspect of a bowling alley. It was painted with lines like P3's track. One of P3's favorite tools—a slideboard—waited unused next to a giant net where I never saw anyone hit a softball off a tee, throw a football, or kick a soccer ball. A tub of volleyballs, golf balls, tennis rackets, and football tees sat quietly.

When Katie Spieler tore her ACL, at some point the therapists handed her off to P3's performance staff, who made her beyond functional. But what about those of us with bigger dreams than getting in and out of cars? Can we get ballistic?

For that matter, what if healthy people came here to become more athletic, and to prevent injuries? I'm reminded that P3 began on a plyometric runway in the back of a physical therapist's office, where Marcus coached Olympians. What could healthy people do in the back of this physical therapist's?

Much of cardiac care now goes to seemingly healthy people. Marcus "guarantees" movement insight is coming to us all; it's a case he has made to athletes, coaches, and general managers (and onstage at Harvard Medical School when his alma mater named him the August Thorndike lecturer). It feels inevitable, but it will take a freight train of dollars. I spend time in P3, watching Jalen Williams and Chet Holmgren enjoy the finest training in the land, then fly home and realize almost no one has access to Eric's force plates or SIMI motion-capture system. We each need workouts personally tailored to our own strengths and weaknesses, but our towns lack not only those force plates, but those brains. Very few universities offer biomechanics as a major. Trainers well steeped in soft-tissue injury prevention are so rare that at P3, they refer to them as "unicorns."

The future Marcus envisions has much less time watching TV or stuck on Zooms, or even working out on the elliptical—and a lot more time surfing, climbing mountains, and jumping into water. When you go to the gym, though, Marcus says, let it be prescriptive to your exact body. Histories and movement patterns carefully assessed. Your MRI in the gym file. The workout hard-core, challenging, fun, and entirely designed and overseen by a trainer with a piece of paper detailing your exact issues—skilled eyes on your body as it moves.

Marcus watched me walk around a few minutes before noting I had the body type of a blender, which in NBA players comes with a 300 percent increased likelihood of lower-back issues. Everyone knows we land with three moving parts—ankle, knee, hip—that can absorb force. The news is that some bodies recruit a fourth, the lower back, and start moving things around there. The injury and the solution are both movements.

A few months after I began physical therapy, I regained the basic abilities to move and swim and even ski. Most of the time, I felt fine. But I ran out of physical therapy visits long before I figured out whether or not I could jump or run. I wanted to be ballistic again.

I texted Marcus—"Can we jump off a cliff?"—and booked a flight.

# 20

## STORK PRESS

*In dreams, anything can be anything, and
everybody can do. We can fly, we can turn
upside down, we can transform into anything.*

—Twyla Tharp

THE DAY I arrived, Marcus had just met with a seventeen-year-old volleyball player. Something was up with a disc in her back. He remembered her saying she was just going to get a little surgery, just to make sure everything was good.

"We don't have surgeries like that," Marcus mused. We were crammed into the Audi on the way home from Keean's Sunday morning soccer game at UCSB. I shared the back seat with my rented road bike. Keean had six-foot German exchange student Lenny strapped to his lap.

Coach Steve Kerr had what was supposed to be routine back surgery after the Warriors won their first title, and he ended up losing spinal fluid. For a time, it was hard for Kerr to stand. The next season, the Warriors set the all-time NBA record by winning seventy-three out of eighty-two games, steamrolling everybody. But that season steamrolled Kerr, too. He missed forty-three of those games with chronic pain. Years later, the headaches still hadn't stopped. He tells everyone to avoid back surgery at all costs.

He's far from alone. There were almost 19 million orthopedic surgeries in 2022 in the United States. "There's very little thought," Marcus says, "about how that happened. And that it might have been preventable."

A historical cohort study in *Spine* found treating lower-back pain with fusion is "associated with significant increase" in disability, opiate use, and prolonged work loss compared to treating similarly injured people with less invasive methods.

According to a systematic literature review in the *American Journal of Neuroradiology*, a huge percentage of people have messed-up spinal MRIs and don't even know it. More than half of pain-free people aged thirty to thirty-nine show up on MRIs with disc degeneration, height loss, or bulging. The researchers suggest that, "even in young adults, degenerative changes may be incidental and not causally related to presenting symptoms."

Among twenty-year-olds who are not complaining of back pain, 37 percent have disc degeneration detectable by an MRI, 30 percent have disc bulge, 29 percent have disc protrusion. "Our study suggests," write the authors, "that imaging findings of degenerative changes . . . are generally part of the normal aging process rather than pathologic processes requiring intervention."

Which means when a doctor shows you an MRI of something wrong in your back, you might or might not need to do anything about it. Marcus's own back would fit those findings. Decades ago, he had a snowboarding accident that broke a little bone off a vertebra that pressed a nerve, hurt like hell, and earned him a trip to the MRI tube. The imaging found a host of issues similar to those affecting the volleyball player. He declined surgery and went decades without hearing from the disc again.

Surgery would presuppose some doctor knows better than the body, but ha! The body, Marcus says, is a *self-healing machine*. He nods to Keean, who is back playing soccer after a broken collarbone. Keean's body, Marcus says, has "invested all these resources making Keean's collarbone stronger than it was before . . . because you've shown for whatever reason that your environment needs more out of your collarbone at that spot."

"That," says Marcus, "is where the body is most brilliant."

We're all a little broken. That's not news. But underappreciated is that, even in the crammed back seat of an Audi, we're simmering with healing brilliance.

Marcus adores renal physiology like my daughter adores Taylor Swift. The kidney monitors and modulates blood pressure, electrolytes, and acidity. It extracts toxins. It produces *erythropoietin*, the artificial version of which was the magical drug that drove Tour de France cheats. In other words, if you're a guy like Marcus who wants to bet on the body's ability to handle whatever you throw at it, you're betting on the kidney. The crux of the research of Horvath and Noakes is that a well-trained body can almost magically adjust to heat, cold, massive workloads, and anything else you throw at it. That intelligent coping comes, largely, from the kidney. "It's so complex!" says Marcus, turning into the driveway.

Marcus suggests that the body's wondrous ability to achieve a steady state applies to musculoskeletal issues, too. If you know how, you can move out of the pain cave. P3 has shone the flashlight around the dark recesses of our ignorance, but the big finding isn't how much P3 knows; rather, it's *how much more* the body knows. The gist of this Audi trip is that we benefit from biological brilliance we are only beginning to understand. Squid and owls move in ways that seem magical—using DNA and nerves that look a lot like ours. *We* might be magical, too.

A few hours later, Nadine puts out brie and a nice local cheddar, banana bread, grapes, a long, warm crusty loaf sliced into chunks, and three salads made from the haul from Kira's job at the farmer's market.

The conversation turns to dreams. One guest says she has a recurring nightmare that a big black spider drops from the ceiling and crawls under the covers. She throws off the blankets and wakes up her husband. Jonas has a recurring dream that he wanders to the back of his property and finds a barn he hadn't known was there, and it's full of old cars.

"Here's this thing we spend a third of our lives doing," Marcus says, "and other than to say sleep is important, we really don't understand it." We touch the cosmic on the wings of nighttime neurophysics, and shrug off the crazy in the morning.

"If you don't sleep for five days," Marcus points out, "you die." In his Boston days, Marcus was dazzled by the impossibility of working through two straight nights. He had a misogi-type urge to persist, but says, "You cannot. Not with coffee, not with willpower." Marcus was impressed by the body's determination to steady his state. And by how little anyone truly understood why, even in one of the world's finest hospitals.

Dreams, he points out, are an even tougher nut to crack. They feel absolutely real. We put our bodies through these little misogis, feeling the full terror of a spider—for *something*. Probably, it's evolutionary genius, but the point is unclear. All we know is that the unknown parts of our human systems are often surprisingly useful.

Marcus's recurring dream is that he's free-falling off a cliff. He feels all of the terror of impending death as the ground rushes at him until . . . he sticks out his arms and begins to fly. Marcus asserts that this is a normal dream. A wine-and-cheese party worth of people disagree. Marcus says he thought it was normal because Mila once said she had the same dream. Mila nods, says that's true. And it's hard to know what's crazier: thinking we are supernatural or thinking we are not.

After my moaning winter of lumbarmageddon, I'd been seeing the world as an owl in a box. And I had journalistic ethics that had me telling people I didn't want to become a patient at the Lab. Marcus essentially said, "Enough already" and arranged a visit.

The Lab's CEO, Alex Ash, studied my medical files, then kicked everyone out of the Lab's kitchen for the most thorough medical interview of my life—he got me to remember things I had forgotten about high school injuries, and beyond. Then he set me up with Jeff Rosenthal and Adrina Lazar, who gave me Arsenal-red, leave-nothing-to-the-imagination pajamas for a biomechanical assessment.

After a warm-up, the goniometering of my hip angles, ankle flexibility tests, and a clipboard's worth of notes, we walked a block to P3. There, I was slathered in reflectors for a half hour of stepping off boxes and jumping, hopping off one leg and landing on that same leg on the force plate, explosive one-off skaters, assessing thoracic mobility with a big pole across my shoulders, and more to produce the computer's view of my body.

We spent a few minutes looking at the screen, noting with sunny tones that I had excellent shoulder mobility, and really stuck the landing hopping onto my *left* foot. On the right? We'd talk more after they'd processed all the data.

All of that made me late for a physical therapy assessment, so I hustled back to the Lab to find an empty table in front of physical therapist Zachary Finer. In 2014 his dad read about P3 in *Sports Illustrated* and left Chris Ballard's article on the kitchen table for his high school–aged son. Zach read it and made working there his life's mission. When Zach saw the words ERIC LEIDERSDORF on a name tag at the World Biomechanics Congress in Dublin, he felt like he was meeting a celebrity.

As a PT, Zach's job was to offer an independent second opinion of my biomechanical situation. He poked, prodded, and reached freakishly similar conclusions as the higher-tech assessment.

Especially: a curious lack of strength, coordination, and range of motion in my right hip. As I lay on my side with my knees opened in a clamshell, Zach pushed hard on my upper knee and told me to resist. I fought him off equally hard on each side, I felt—but when he got to the right side, Zach almost cracked up at how easy it was to move my knee. "See?" he asked.

Zach used the remainder of his time to dig fingers into the psoas on the front of my hip. I felt he had found the physical location of my anxiety. "You seemed," I said, "like such a nice guy."

Then it was time for my first workout. Every P3 workout begins with the words "get on the track" for P3's signature warm-up. More than anything, Marcus wants workouts targeted to the individual, but it occurs to me that these thirteen steps of warm-up are special, a gem: a session packed with P3 insight but designed for everyone. It's something of a calling card. If I forgot one of the moves, every customer, staffer, the receptionist, or former NBA center Jahlil Okafor (who happened to be in Santa Barbara that day, rehabbing a torn Achilles) could fill in the gap.

HEEL WALK

TOE SLAP

LUNGE TWIST

REAR LUNGE REACH

SIDE LUNGE

INCHWORM

SINGLE LEG ROMANIAN DEADLIFT

QUAD STRETCH

KNEE HUG

LEG SWING (SIDEWAYS)

LEG SWING (FRONT TO BACK)

FIGURE FOUR SQUAT

90/90 ROTATION

CROSSOVERS

First, you do a length of the track walking on your heels, to prepare your lower leg to land with a dorsiflexed foot. Next come toe slaps: take a step that lands loudly, as if you're slapping the palm of your hand loudly on the floor. Then it's a lap of lunging forward with a twist over the forward knee. Then a "rear lunge reach," which is lunging backward with a long arm leading the torso as it tips to the side of the leading leg. Then side lunges, carefully poking that butt back behind you as you press into the landing leg. Then we inchworm: begin standing, then bend over in a hamstring stretch with legs fairly straight. Walk your hands forward into a plank push-up position, then walk your feet up to your hands for another hamstring stretch. Then stand up and start again. Then there's a walking lap of alternating single-leg Romanian dead lifts, a quad stretch where you simply grab your foot behind you and pull, and knee hugs. Then you put a hand on the wall for leg swings sideways and front to back. There's a bit of balancing in the next lap, which is figure-four squats, where you rest an ankle on a knee, then sit down on the other leg.

Then, lie on the floor to get your back rotating. First is a version of an "open book" stretch, where you lie on your left side, left leg straight and right leg bent. The left hand pins the right knee to the floor. Rotate your upper body backward toward the floor, ten times each side, with your gaze following your upper hand. Crossovers are the thirteenth and final step of the warm-up: lie on your back with arms out, raise your right leg straight up into the air above you, and then, with your shoulders pinned in place, sweep that foot across your body to touch it to the floor on the left. Ten times each side.

Later, I ask Marcus why we warm up at all. He says that the research shows warm muscles can handle much bigger loads, maybe twice as big, without tearing. And he says warmth gives muscles, tendons, and ligaments "a little bit more freedom . . . when you start doing higher force movement, you don't run into resistance as quickly."

Then Marcus references the neuromuscular coordination honed by plyometrics and the impulse box—the nervous system signals that make movement like language and language like movement. Singers and actors do tongue twisters and scales backstage to prime their instruments. Bouncing up and down the track, Marcus suggests, similarly attunes the muscular control of more athletic movement. (Marcus also points out that athletes almost all find it helpful to listen to music before a game; he suspects there might be a kind of rehearsal going on, of neurological timing. The rhythm of the music might approximate movement impulses.)

After warm-ups, we used a medieval torture device shaped like a sword with a ball on the end to locate, and then press incredibly hard on, the psoas that Zach had just probed. The psoas tool felt terrible, then wonderful, and I ordered one that day. I spent a minute with each quad on a foam roller, and then lay on the floor with my heels on a box and my quadratus lumborum—or QL, the deep back muscle on either side of the spine at the belly button—pressing into a lacrosse ball. After a minute with each side of that, I flipped into a plank position and sank my weight into the lacrosse ball on the very outside of one hip, pressing into the tensor fascia latae, which held a surprising amount of stress. I have repeated those eight minutes of hip myofascial release a hundred times since—it can change your day.

"What," I asked, "are all those marks on the wall?" A couple of dozen gray arcs, sliver moons a foot or two off the floor, populated the wall along the track. "Oh! You'll see," said Alex.

Meet the heel slide. You can use a sock, paper plate, or towel, but Alex handed me a little bootie, like a surgeon might wear, to slide over my shoe. Lie on your side with your head, shoulders, and hips flat to the wall. Turn the toes of your upper foot toward the ceiling, and press the outside of that foot into the wall as you slide your foot up and down, trying as hard as you can to leave a cool mark like everybody else did. It fires a novel quadrant of the glute.

In my personalized workout, there were too many moves to track. I stood on one leg, passing a kettlebell from my right hand to my left. Katie Spieler did it with her foot on plates; as a beginner, I stood on the floor. I squatted on one leg with a dumbbell in one hand, and while my standing leg was bent, punched the weight diagonally down across my waist. I pogoed like Kyle Korver.

Then came an exercise that has long been routine for me: single-leg Romanian dead lifts, just like in the warm-up, but now with weight. The movement is basic: stand on one foot while you hinge forward at the hips, your upper body goes toward the floor as your trailing leg reaches out behind you. I have done this 10,000 times, in hotel gyms, at home, on the lawn, on the beach. I am not sure I completed three reps before Alex noticed I fired the wrong glute.

My movement epiphany probably sounds silly. Most people know which glute to use—the one on the standing leg. But somehow, I had been balancing using the wrong muscle. Hips want mobility and stability; this move wants each hip to pick one, and I flipped them. My misunderstanding of my own hip muscles explained much of my sloppy movement, every step I had ever run, every stair climbed, anything I had ever done on one leg.

Jeff and Adrina pointed out the video of my drop jump. As my right foot headed for the floor, it towed my right hip down with it. "See that pelvic sway?" Yes, 100 percent I did, Jeff. If my pelvis formed the cup of a wineglass, I was spilling Barolo. This hip instability is the natural consequence of the glute of your standing leg doing no work.

I tried to keep my pelvis together doing a length of the track in Romanian dead lifts, hinging forward with a kettlebell.

Then Alex had me march the track pressing a plate overhead so my biceps pressed to my ears. With each step I was to raise up on tippy-toes, while using my psoas to pull the other knee high up front. That day, there wasn't a single step where I managed to get the plate high, my bicep by my ear, knee pulled high, foot on tippy-toe, and the glute of the standing leg firing. Alex would note whatever was missing and, as often as not, when I fixed that, I'd lose my balance and put a foot down.

To work on hip range of motion, I lay on my back with my left sole on the wall, and my right ankle across my bent left knee in a version of the classic supine figure-four stretch. The trick was to push the right knee outward with the right arm, coaxing the piriformis, glute med, and others into opening up. But after ten seconds or so, Alex had me fight back with the knee, pushing it hard into my hand. The antagonist struggle helps the agonist muscles relax—immediately after the strain, I resumed pushing on the same right knee, in the same position, and now the knee magically dropped an inch or two further open. We repeated this a few times, and soon I could easily see the sole of my own foot for the first time in decades.

Then I met my nemesis: the stork press. I stood on my left leg, with my right knee pulled high in front. My right hand held a fifteen-pound medicine ball above my shoulder, like a waiter with a tray. The trick, then, was to push the ball to the sky, to put my bicep back in my ear.

The work of a push-up is as familiar as classic rock. The stork press felt like new jazz, or a shoulder press on a slackline. The sweat rushing from my temples wasn't from the weight of pushing fifteen pounds, it was from the weight of pushing fifty years old while mastering a new dance step. But I persisted: my future of free movement required it.

The balancing act of the stork press plays out diagonally through your core, between your still left foot and weight moving up and down in your right hand. The angle seesaws through different muscles as the weight climbs and falls. I had moments of overthinking, moments of brute force, and moments of listing. Alex asked which glute I had working. Ten reps on my left side left me emotionally vulnerable; ten reps on the right caused me to pitch and stumble so badly, I worried I might hurt other Lab clients.

Alex suggested I complete my stork presses with a wall to my right. That quite obviously kept me from tipping over, but more subtly made pushing the weight far easier. It felt like the energy I had been using to balance and think could now be applied to moving, which is the point.

STORK PRESS

Over the week, I visited the Lab three times. Each time we did the same warm-up, then a distinct workout. They all fit the same pattern: sets of three or four exercises organized into complexes. The first complex was the warm-up plus some soft-tissue release and stretching. I'd do that once. The second complex I'd go through twice, and then the third and fourth complexes I'd go through three times each. Near the end, there was always a signature challenge along the lines of the stork press, a weighted side plank, or a hip bridge with a heavy medicine ball squeezed between the knees and toes elevated.

They'd give me these three workouts in their app, with little videos to demonstrate, and I'd learn them all. There was so much optimistic talk about my future of movement, swimming in oceans, running up trails, zipping around on bikes . . . nothing would happen in a day, but a lot was possible in a month—and everything in a year.

Katie Spieler doesn't go to P3 anymore. But she's a bigger part of the Elliotts' lives than ever because most of the Elliott children train at the East Beach Volleyball Academy. They go there because they love volleyball and because they love Katie. Katie has a superpower the Elliotts hope will rub off: Katie, Marcus says, is "frictionless," in pivoting to the next thing that must be done.

Tearing her ACL gave Katie a new habit: she swims in the ocean every day. At first, her sister had to carry Katie and her battered knee out through the waves. The Pacific's too cold for Marcus without a wetsuit, but Katie says she and her friends never wear them. "We just do it for the love of being in the water." They see dolphins and seals and pelicans and read news stories about great white sharks. Katie has seen orcas and humpback whales from her dad's boat. "They've gone under our boat, gotten really close."

Does that make her hesitant to jump in? "I really have no fear of the ocean," says Katie.

Marcus says that for Katie, "work has no emotional baggage."

That's the goal. I walked out of my first day at the Lab amazed at my ineptitude but exhilarated to be on a journey to discover the new movements I'd need the second half of my life.

I believed I could address whatever was wrong, but what was it exactly? Was I someone with an unstable right hip, or was I like that NBA rookie who wouldn't put his heel down, guarding against using my right side? I'd have to be frictionless in exploring that, too. The next morning, Marcus invited us on a hike.

# 21

····························

## SEVEN FALLS

····························

*I will feel a whole lot better when I hear the frogs.*

—E. B. White

On Saturday's hike, I tried not to think about how I placed each foot. But when you're fresh off six months of the worst pain of your life, a day past watching video of your effed-up movement, and accompanied by a doctor who advises you to relax . . . "natural" is hard to define.

At that very moment, Adrina was at a computer, working the weekend processing the persnickety data of my assessment. On Monday, I was due to stare at a screen and drink in the full lessons, numbers about my hips, back, and feet from machine learning and servers full of data. It's hard to imagine how Marcus could have pioneered a more invasive system. P3 has data-based opinions on how you place your toes as you squat, how your head sits over your neck, and how you breathe.

But Marcus says the trick is for athletes to develop new movement tools in the Lab or at P3, and then just trust their genius bodies to move well when they're being athletic. The body, he says, tends not to leave great tools unused. Leave the key fob to a new Porsche lying around, and someone will slip into that driver's seat. So, I moved smoothly, naturally, as thoughtlessly as possible, only sometimes fixating on the glute of my standing leg as we nudged upward through the three winding miles above the water plant, on our way to Inspiration Point.

Paul Dudley White might have baffled people with the nuances of his echocardiogram findings. But the takeaways were dead simple. To prevent heart attacks, the recommendations are, roughly: move your body, don't smoke, skip the steak, eat a salad.

In the game of preventing musculoskeletal injuries, the obvious early wins come from strengthening unstable hips, stretching immobile hips, and landing toes up. These are P3's biomechanical commandments. But another one is: *Get after it*. Marcus is always punctuating texts with that fist emoji. He believes we can all move in big ways.

Hikers generally walk an hour or two to see the views of Inspiration Point, but Marcus had ideas we'd keep going down the other side. As we passed Inspiration, I asked Marcus why we, as humans, evolved to jump.

I thought he may have insight from evolution or physiology. Instead, he almost couldn't see it as a legitimate question. "What animal can't jump?" he countered. "You have to land," Marcus continued, "whether you play in the NBA or you just occasionally do a little jog or chase down your dog or something."

That's the part that seems to me like it might hurt. It's also, Marcus suggested, the most important thing to get right. Much is missing, Marcus explained, if we train just to build muscle. "When you get this thing strong," he said, pointing to his thigh, "okay, it's strong. But what's that say about how you're going to move? It doesn't. It really doesn't say much at all." Much better is to train your landings, to think about how you interact with the ground, and, Marcus says, "evoking this crazy supercomputer" of our movement systems.

He gestured to a creek bed, lumpy with rocks below. No robot on earth can yet run on terrain like that. There's just too much to process. But, as a measure of the competence we carry with us everywhere, humans can. A study in India had serious runners navigate a bumpy track strewn with obstacles, and found they could run just about as fast as on flat ground, in many cases without even looking down.

"Think about, like, how much timing has to be impeccable!" Marcus said, excited now. "There's so much feel involved in this. You know, initial mechanics have to be set up or you're just behind the eight ball all the way through it. And those are crazy complex systems."

When we reached the stream, we paused. Marcus was talking about how the water was so clean, you could probably drink it. A frog plummeted in like a child from a fig tree. Duncan said when he was little, seeing a frog like that made his whole day. "There's something just so lovable," Marcus says, "about their little movements." I asked if I should emulate the frog and jump in, too. Marcus winced. "Well," he slowed down, picking his words, "let's go up higher. Okay?"

What came next was an ascent of the river, hopping rock to rock. With a little skill, you can keep your shoes dry. I hadn't jumped in months, other than in the P3 assessment on the soft gym floor. We'd see how my back liked one-footed landings. But I told myself robots could not follow us and Marcus was a doctor who had seen my chart and declared me ready for this.

The bendy rubber sole of my left shoe curved across warm rock. To hell with it. Press into that base, with an active glute, and throw my delicate right leg out over the cold, clear water. The bet was that I could fly perhaps a yard—maybe not exactly sixteen-year-old Zion forces, but ballistic anyway. The ball of my right foot found a giant black stone, my lower leg worked a tad like Elon's landing rocket, my heel didn't slap down, and my sole had an open line of communication to my right glute, which kept my hips from swaying.

Nothing hurt. Not even a little.

"Oh, hell yeah," I said, maybe out loud. They say when you die, your whole past flashes before your eyes. When I landed, it was my future that flashed by. I felt free as an owl climbing out of a cardboard box. *Let's go.*

Before long, we scaled a steep and reddish riverbank. "These are amazing rocks!" said Marcus. "It feels so good." In places, it was almost like the climbing wall at the gym with the dead-easy handholds—and with a breeze and a view and a warm piece of ancient solid nature in your hand. "Beautiful," Marcus narrated. "Like, very primate."

We pop over the lip and into a Georgia O'Keeffe masterpiece at the bottom of a towering canyon, a clean line of waterfall cutting deep between huddled towers of silver-brown-gray moonscape.

"I don't want to speak too early," says Marcus, "but I feel like we may be the only people up here."

We are standing on the dry, high right end of the mouth of the canyon, a curved grin of rock, with a smooth quarter inch of water racing across to form the creek we have been tracking. Above that rim, though, is an idyllic pool, just plunked there to make life amazing, fed by a waterfall plummeting five or six feet from another smile-shaped lip. The place where I thought of swimming earlier seems so silly now.

"This water is water that's landed up on top of this mountain, three weeks ago, four weeks ago," says Marcus. It is clearer than clear. "It takes that long to trickle through all the rocks to eventually end up down here, perfectly pure. Purified, right? Isn't that amazing?"

A tad high on the idea I'll get to move freely the rest of my life, I wonder if this is the most beautiful place I have ever been. This little heaven pool is the bottom, somehow, in a chain of seven. The place is called Seven Falls. Marcus says we should leave our shoes, and if you cling to the rocks high and to the left, you can sneak past the first couple of pools and it's pretty nice up there.

I barely listen. Shirt and shoes abandoned, I initiate my own Pentecostal baptism, wade in, and stand awhile. Marcus and Duncan chat, but I have sun on my skin, the fizz of waterfall in my ears, and therapeutically cold water on everything belly button down. Later, they will make fun of me for how long I simply gaze upstream. When they yell, I turn, and Marcus throws something; I catch a single Trader Joe's chocolate-covered almond. I wave thank you and eat it. Marcus reaches into his shorts pocket and tosses another; delicious. Then I turn back to my main project: examining the rock to the right of the waterfall.

The longer I look, the less clear it is how I can climb that thing. Finally, I just swim over there, stick my right hand in the one good handhold, and feel around with my left hand for anything up top that might work. As I haul myself up, chest down like a salamander, I see

Marcus and Duncan creeping out onto the rock face twenty feet up to my left. Marcus picks handholds and footholds, Duncan follows tidily.

I prefer my route. But the second waterfall is taller, the rock as slick as an orca, the water below treacherously shallow.

Now I want to switch to Marcus's high road, but it's a route of slight hand and footholds, and I am soaked. I climb halfway up, testing the grip of wet feet; it's absurdly slippery. Shoot. I pause to let the sun dry me a bit.

The simmer of prefrontal concern kicks on. I daredeviled plenty as a kid, but suddenly I remember the chicken times. I didn't always go off the high dive or bike along the log. Maybe I have reached my destination? Am I done? Am I already in a perfect spot? Can I be satisfied in the second pool?

I look upstream and see Duncan small in the distance, arms and legs carrying him smoothly up toward the horizon. In the foreground, Marcus stands upright as an owl. His face betrays nothing, eyes library calm. But he has intent, his bare left foot placed before his right, like he might shoot a free throw. He is up to something.

A few weeks earlier, Marcus and his whole family hopped a fence to trespass on the six hundred acres where he grew up. Now, it's part of the multibillion-dollar Duckhorn wine conglomerate. Marcus showed his kids where he swam with horses, and where the house stood until it burned, and where the rattlesnakes sunned themselves.

A lot of what Marcus cared about is gone, but not the fig tree where he learned to fly. He has jumped off bridges, rocks, and a cliff so high in South Africa that his friends intervened, trying to talk him out of it. We learn a lot more about our bodies from big ballistic movements, and if we can relax, we can tap into the intricate genius of ancient motor neurons. Marcus and Mila brought home six cuttings of that fig tree, which are rooting in pots, and soon they'll grow all over the Elliott property for another generation to climb.

I won't forget the sun on Marcus's face as he presses his dry toes into rock. He takes to the air with the calm delight of a gardener finding a nice tomato—then disappears into a gulping sound below.

Fuck. That's where I need to be. Not here. Just zen it. Less thinking about hitting your fingers, Marco. Wiping wrinkled white toes on dry spots before jamming them onto narrow ledges, I sidle out onto the face.

Where the face curves left, there's an impressive drop below. I'll have to scoot around a little outcropping, like an NBA defender navigating a screen. The problem is the surface. The handholds are pointless fingernail cracks. My right big toe is jammed into the only good hole. My next move will be swinging my left foot forward . . . to where?

I press my wet chest into the warm rock and do exactly what Marcus wouldn't recommend: I think. There's nothing to grab, but there is a decent foothold up around the corner. Ideally, I'd have my left foot anchored beneath me, then I'd keep my chest, hands, and eyes on the rock as I wrap my body around the outcropping and reach a right foot forward.

But I have to live my wrong-foot reality. And there's a way it can work. It's a big bet. I put all my weight on my right foot, and then slide my left foot between my body and the wall. Then I have to get long, so I turn my back to the rock. *Woop!* I reach my left foot up around the corner. Then I entrust my full weight to that foot's damp outer edge. It's what I have, it's what I do, and it works.

From there I dash easily up the next rise, to Marcus's spot above the third waterfall.

The jumps I had done in the assessment were my first ballistic moments in months. This hike, these rock hops, have me more certain than ever that I am bouncy, that my hip muscles are okay, that my hip weakness is borne of fear, like the NBA player who won't put his foot down. Jumping is my way to goose it up, to tell my brain, *Thank you for your concern, but I got this.*

Looking downhill, the water rushes to our left. Marcus has scrambled out, wet and smiley, and points to the pool below, frothy white with water. "I just jump there," Marcus says, pointing to a clear part, "because you can see the obstacles." There are none.

And so, I just go. Full throttle, feet first, deep into a tumble of bubbles and drinkable frigid water. I arrive back to my sunglasses, hat, and Marcus with a grin, fresher than when we left the car, hours of walking ago—or at any time since the prior summer.

Marcus is excited. He wants to talk about something that happened a few minutes before. Duncan, he says, made some decisive moves—*boom, boom, boom*—across the rock, and from down in the water I hollered "Yeah! You got it, Duncan!" We are all trying to give each other the confidence to move.

As we warm ourselves in the sun, we talk about kids—Mila's new school, my daughter, Molly, who was, at that very moment, touching off a career in sports. "I have a picture of Keean up here," Marcus says, "when he is like a year and a half or two years old. Literally, like, he climbed his little body all the way up here."

Kira has just launched seriously into volleyball. I asked if Marcus is worried about her ACLs, an epidemic in her cohort. "I would be crushed at such a deep level," says Marcus, actual palm to heart, "if my daughter had a non-contact ACL."

But he started giving her little physical challenges when she was five. He knows Kira's movement. "You want to absorb force through your hips. When you can't absorb through the hips, that puts force more through knees and ankles. Mechanically, you don't want big valgus, you don't want big rotation. You want neural timing so it's really linear: hip, knee, ankle, really in line. Kira can do that. If you watch her single leg bound along, I think we're in a good place," Marcus says. "Kira's got a buffer, an injury buffer. We can provide that for all these kids. It doesn't take a lot of tech. That so many people still get hurt is a real pain point for me."

Duncan scrambles down to us. He and Marcus pick down the way they came, but I descend in the water like a frog. I've been worried about my back, my poor conditioning, and a hundred other things. At the bottom, I have functioning hips and feet, ice water on my skin, a candle in my chest, and a flicker of invincibility.

# 22

......................................

## INCREMENTS

......................................

*Nothing happens until something moves.*

—Albert Einstein

THERE'S A LOT of money for doctors who provide "immediate and visible success," like cracking open a patient to unblock an aorta. Cardiac surgeons are well paid, writes Atul Gawande in *The New Yorker*. You know who makes about half as much? Cardiologists who do the work Paul Dudley White recommended, who tell their patients to ride their bicycles, thinking long-term to prevent aortas from getting blocked in the first place.

"The pattern," Gawande writes, "is the same everywhere." We know prevention is cheaper, but we raid funds for tough-to-appreciate road maintenance to pay for exciting and noticeable new bridges. "Does anyone reward politicians for a bridge that doesn't crumble?"

Gawande writes that medicine is decades into this debate, and gestures at people he calls incrementalists. "The patterns are becoming more susceptible to empiricism—to a science of surveillance, analysis, and iterative correction. The incrementalists are overtaking the rescuers." Gawande's conclusion is that, instead of body shops, patching people up after crashes, "we all need to be pit crews now."

I thought about all the good people who had tried to cure my back. The primary care physician, the pain specialist, the massage therapists, the chiropractor, the physical therapist, the trainer, the sports medicine doctor. Everyone smart and eager to help, working away for months. But no one had a force plate.

At P3, I found a pit crew. Alex ran the numbers and showed me that in the drop jump, my toes hit the ground like Katie Spieler's, which brought my full weight down through my heels. "So, this is six and a half times your body weight crashing down into the ground," Alex said. Alex said the "threshold for where there becomes a lot greater injury risk is at about four and a half."

And the force plates had a note on my symmetry. I didn't have any: a 30 percent asymmetry during the braking phase and 20 percent asymmetry during the push-off phase. Alex wanted my asymmetry "for sure" below 15 percent, ideally below 10. I was wrapped in red flags that only P3 could see. How long is the right amount of time to live without these fundamental measures?

Another mystery lingered: What caused this instability in the first place? Alex said "there's probably a lot of guarding," and it'd be interesting to see how I looked "with, like, being less afraid to move." His best-guess theory of the case: Born with hip dysplasia, I had concocted compensation patterns which had mostly worked. But I had underdeployed my glutes and overtaxed other soft tissues in the neighborhood.

When Marcus left home with three hundred dollars and a mission, he was in a wicked pinball game. He bounced off the molecular biology and physiology of Horvath as an undergrad, then modeling, triathlons, medical school, and Tim Noakes—all without finding anyone who knew much about preventing injuries. Bert Zarins, the New England Patriots, the Seattle Mariners, and the force plates moved him a little closer. But only in his forties did the little metal ball of Marcus's life story pop into the slot marked *motion capture*. Even then, merely finding the right dataset wasn't enough, because most of the data was confusing misdirection. But several thousand hours later, while basking in the insight of P3's athletes, physical therapists, doctors, and trainers—and after Eric went back to school for a data science degree, with a sidecar of machine learning—P3 had begun finding signal in the noise.

How do we get that actionable insight into more hips and feet? In the *Atlantic*, Derek Thompson writes about the history of disease prevention. "Progress is our escape from the status quo of suffering, our ejection seat from history—it is the less common story of how our inventions and institutions reduce disease, poverty, pain, and violence while expanding freedom, happiness, and empowerment."

Edward Jenner injected his gardener's son with fluid from a diseased cow. When it worked to prevent the smallpox that had ravaged the globe for decades, Jenner came up with the word *vaccine*, named after the Latin *vacca*, for "cow." And the idea is that, after Jenner's miracle, smallpox prevention encircled the globe on its own.

But in reality, Thompson argues, it's as much about implementation as invention. Paul Dudley White's lifesaving heart attack insight lingered for decades in studies and textbooks, and only really caught on with the public catastrophe of a president nearly dying in office. How many people died needlessly in the interim?

"The way individuals and institutions take an idea from one to 1 billion," Thompson writes, "is the story of how the world really changes." The smallpox vaccine, he adds, shot around the globe only once a few kings, several doctors, and a pope vouched for it; Spanish doctors shared it across the Atlantic *in the bodies of live orphan boys*; an energetic young WHO official invented new and better ways to choose whom to vaccinate; and a microbiologist devised a bifurcated needle that dropped the cost to ten cents a patient.

Marcus is betting on such evolution, but it's not obvious what form the movement might take. Marcus is an MD who spent his early years deep in physiology and athletic training. The terms he uses most commonly now, though, are from neuroscience and physics. Eric had a degree in biomechanics, but after working his way to the best chair in the business, he went back to school for data science. Half of P3's employees are certified as trainers or coaches. One studied psychology, many physical therapy. Meanwhile, my high school–aged son asked Marcus what he should study in college to get into the field, and Marcus found the question hard to answer.

Kinesiology, he said, was "too soft." He said the field needed scientific rigor, and suggested that the savviest approach might be to major in mechanical engineering, to get well versed in how things move in general, and then to steer that learning to the body. Whatever the exact title, the rough goal is to combine engineering and medicine, which means the better future we all want will spread on the wings of the incredibly educated.

On my last evening in Santa Barbara, Marcus pulled the Audi to the side of the road near a bridge across Rattlesnake Canyon's Mission Creek. One of his favorite places is named for his mortal enemy, but they say it's because of the way the canyon snakes through the hills. A dozen cars flanked the dusty gravel road as Santa Barbarians enjoyed the trail up into the hills. We plucked our bare soles across shabby pavement, careful stepping, arms out like toddlers. Then a descent to our chosen terrain: rock tops.

Soon, we'd be in the evening shade; Marcus brought a hoodie. But the sun had some angles, and as the outing began, my two feet wrapped a lovely rounded sun-warmed knob of stone. Marcus noted the shapes the rocks demanded from our feet, which he called "so good." On flat ground all day, our feet grow rigid and immobile, which can make you unstable in old age. A man who has dedicated his career to movement, landing squarely on ground contacts as a critical area of focus, was about to lead a master class.

The Elliott family calls this Mila's hike, because it's her favorite—literally child's play. The rocks were close enough together that you can get a rhythm of just walking, right foot on this rock, left foot on that one, over the water and up the stream, with minimal planning. Now and then the course-picking took a foray to the bank, over a fallen oak, or along a rock shelf.

I followed Marcus, but not exactly. We move differently. He's comfortable with rock-climbing holds and avoids cold water; I was sweaty a few minutes in and would rather stay low lest I land hard and excite my hip. And anyway, it felt ancient and natural to ramble along like a pack, on similar but not identical paths.

Here and there, though, a real jump, with a barefoot landing on solid rock.

Two things can happen in your head: You can imagine the pain of missing and freeze. Or you can zen it. Which meant, for me in that moment, being peregrine-still in the eyes, normal in breathing, and springy. It may sound strange, but my body just *knew* how to get there. *Bop*, next rock.

Mila's hike offered an echo of the impulse box. It may seem safest to do one jump at a time, but there's power in alacrity, and stopping is difficult. Faced with a narrow, low-angled chunk of rock face on the left, a small but probably landable rock in the middle of the stream, and then a pebbly patch of beach, for instance, it was tough to see any way to make it one jump at a time—the two middle spots fit one glancing foot better than two standing ones. Springboking through a *bop-bop-bop-bop* is fun, doable, arguably safer. Sometimes, just to prove I could, I took big landings on my iffy right leg.

For a stretch, almost all the stones were spheres—basketball- to minivan-sized—speckled like rainbow trout. My eyes loved Mission Creek, my feet loved the rocks, and my hips and back didn't quibble.

In mountain biking, they call them *skinnies*, normally a log or board that's long and only a few inches wide—the lowest-budget bridge. It's never been my favorite kind of obstacle. Some way up the Mission Creek, eight feet over a splashy rock bowl, a fallen log was the simplest way. Marcus went first, I zenned along after. As he pulled himself up alongside the small waterfall at the far end, he looked over a shoulder and said that if I really had hip instability, I would not be crossing logs like that.

Soon I'd make a personal religion of my P3 Lab workouts. There's a thirty-pound medicine ball in a cubby by the front door for stork pressing and a subtle heel-slide mark on our dining room wall. Little competencies resume: My hips stay level going down the stairs. My dorsiflexed foot and glutes have begun talking; I have my toes up and the landing force going to my glutes, and it's mind-blowing how much easier it is now to jump rope. I feel like I know a secret.

For a period, I found it strangely helpful to run loops, barefoot, through a grassy park—really feeling the toes-up landing. Then I moved to harder surfaces and awakened some Achilles tendinopathy. Alex advised pausing the harshest landings for a period of gradual strengthening, mostly through slow single- and double-leg calf raises. But he added that "ultimately, you really want to get back to loading it again." Prepare to move, and move.

One spring morning I went for a shirtless run on a hill over a sleepy little town on the Delaware River. The sunrise kissed a million treetops. On the edge of a grassy field—it looked like Winnie-the-Pooh's Hundred Acre Wood—I sensed movement.

A large black bear, healthy as the first one I'd seen, but taller, ambled through the shadow. We were headed for precisely the same corner; I was *running* there. When confronting this kind of wild animal, the experts say, your safest best bet is to act like Brandi Chastain in the penalty box: be big and loud and act like you own the joint. I was ready.

Then he swung his tree trunk of a neck, aimed his psychologically weaponized eyes to my face, and boiled my innards. The bear and I both knew I could die. One of us probably knew how my flesh would taste.

Time for my move. Still running, I got big, held his gaze, and lifted my arms over my head like I had won a race. Big, jubilant hands, neck relaxed, chest broad. So strong. That's what I sent out into the air between us, praying it would land.

The bear turned his whole head away, back in the direction he had been walking, and . . . broke into a run.

*HOT. DAMN.* I grinned through intoxicating invincibility.

There were so many firsts. First day back at the gym, first bike ride, first pull-ups. I felt comfortable through the mile run, a hundred pull-ups, two hundred push-ups, three hundred air squats, and a closing mile run of the traditional Memorial Day workout known as "The Murph." I surfed and frolicked in the ocean. I'd run barefoot through downpours on a soleus, post tib, and gastroc that felt like new tires. The fat electrical cables of my nervous system had carried bad news down my leg; now they carry good news up.

"That's exactly where you want to get," Marcus says. "You want to cede voluntary control of those things." The feet, the glutes . . . "they're already communicating."

"At this stage, better is going to be more important than fast," says Marcus. "Use your instincts. Let feel be your driver."

Marcus knows a guy who, after insane amounts of supplements and coaching, could dunk a basketball in his sixties. It got him thinking about aging. "Your body doesn't want to land from that jump," Marcus says, "much less execute it." To be an athlete aging gracefully, Marcus says, the game is to figure out which things are like dunking, which you should simply let go, and which ones are like, say, declining hip mobility, which "you have to absolutely rage against."

Eventually, Rattlesnake Canyon widens, and a waterfall plunges to a super-clear pool. A tree has fallen, so its trunk has become a skinny bridge, and a perfect launch point into the swimming hole.

"That's where I jumped in with Mila last time we came here," Marcus says. It's her favorite thing. But the air's getting cooler and the water is cold.

Marcus asks what I think.

I think we should jump in. He nods. "I told Mila that the rule is if there's water, you get in it. It's always better."

Next thing you know, we are in our underwear, starting onto the log.

"What," I ask, "is the order of events?"

I hope he'll point to where the water is deep, or offer a nod to safety.

"It seems," Marcus says, already halfway out onto the log, "like a pretty simple order of events."

Low sun strafes treetops. In the forest shade, I remove my glasses, grow blinder, and balance-beam out onto the skinny. Marcus points to a spot that looks like the deepest, and says I should jump next to him. Then he says, "Jump shallow!" We count down three, two, one, and rip out through the air.

# Acknowledgments

NONE OF THIS would have happened if Marcus didn't zen it through a project that I know, at times, made him nervous. Any question, any topic, any time, he never failed to be gracious, forthright, and enlightening. Also: almost every substantive talk we had came while biking or hiking up the Santa Ynez. I can smell those stunning mountains right now.

I feel bad for everyone who doesn't have Susan Canavan as their agent and Tom Mayer as their editor—both shaped this book in profound and fundamental ways. Also, thanks to Norton's Rebecca Homiski, Don Rifkin, Nneoma Amadi-obi, Will Scarlett, Meredith McGinnis, Anne Somlyo, Derek Thornton, Julia Druskin, Jessica Friedman, and Lloyd Davis, as well as Bill Morrison of WordCo Indexing Services.

My spouse, Jessica, is a full-on superhero, for everything to make this possible, up to and including editing every sentence at least once, sometimes at 5 a.m. before work.

Molly, Duncan, Jumbly, Judy, and John all gamely joined me in doing P3 workouts and listened to me talk about soft tissues for years. My parents, Charlie and Betty: I'll rope you in.

Louisa Thomas volunteered to be an early reader, then read carefully and delivered brilliant and profound feedback that led to a total rewrite and a far better book. That was a great kindness. A hardy circle of close friends and family read photocopied early drafts and taught me a lot with their comments.

Travis Moran offered many wonderful edits, including, importantly, rearranging the first few lines of the book, and clutch insight into subtitles.

Nadine, Keean, Kira, and Mila hosted me, fed me, and let me steal Marcus's time.

Life is so fun when John Early illustrates it.

The illustration of Katie Spieler on page 73 is based on Mpu Dinani's wonderful photograph.

I don't know much that matters about basketball that I didn't learn from my business partner David Thorpe, who joined Jarod Hector and Travis Moran in covering for me at TrueHoop during long book absences. Randy Shain trusted me and bought me time.

For several months of the writing of this book, I was wholly disabled, and probably still would be, except for the extraordinary brilliance, compassion, wisdom, and hard work of Nicole Domanski, Jennifer Mintz, and Sameer Siddiqi. And at the Lab: Alex Ash, Jeff Rosenthal, Adrina Lazar, Zachary Finer, Kaweena Warren, and Jeff King changed my life.

Eric Leidersdorf is a saint and a scholar. I've lost count of the number of times I called or texted him, entirely confused, and he never made fun of me once as he patiently shared his uncommonly valuable knowledge—even in the first weeks after he became a dad.

Leah Borkan answered all of my dumb questions, and even reviewed some of the trickier biomechanical passages for accuracy.

Adam Hewitt has a very hard job, but does it while remembering an incredible number of names and always having a smile and a good word. I learned a ton from Tom, Jon, Jack, Nick, Trent, Bram, Tyler, Jake, Eddie, and everyone else at P3. Katie Spieler, Rachel Zoffness, Steve Magness—so many interviews for this book were inspiring.

A guy named Brad Stenger first said the words *Marcus Elliott* to me. Chris Ballard wrote a wonderful P3 story in *Sports Illustrated*. But it was working with Tom Haberstroh and Baxter Holmes that really put P3 front and center in my mind, and I'm so thankful for that.

# Notes

THE INSIGHT IN *Ballistic* was collected from sources as noted below, plus 104 interviews conducted between February 2022 through June 2024. Forty-one of those interviews were with Marcus Elliott, mostly while riding bikes uphill above Santa Barbara (downhill it was too windy). Twelve interviews were with Eric Leidersdorf, and twenty-three more were recorded, largely in person over four trips to Santa Barbara and two trips to Atlanta, with one or more P3 staffers, including Adam Hewitt, Leah Borkan, Alex Ash, Jon Flake, Jack Armstrong, Trent Reeves, Tom FitzSimons, Eddie Dimas, Nick Gibson, and Bram Krieger. Nadine Elliott and Jonas Jungblut spoke in person in Santa Barbara and by phone. Matt Osborn was interviewed over two phone calls and one meeting. Stein Metzger was interviewed in person and by phone. Dave "Boot" Bond and Charles Bethea both spoke by phone. Experts Rachel Zoffness, Steve Magness, Charles Kenyon, and Mark Briffa were all reached by phone. P3 athletes, including Kyle Korver, Tichyque Musaka, Amber Melgoza, Tacko Fall, Kim Yeon-koung, Jabari Walker, Cole Swider, and Benn Mathurin were interviewed in person. Katie Spieler was interviewed over three phone calls. Experts on sports training interviewed by phone include Mark McKown, Jeremy Bettle, Charlie Torres, and Justin Zormelo. Wally Blase of Fusionetics was interviewed in person in Georgia, and Phil Wagner at Sparta Science in Palo Alto. A number of NBA sources—players, agents, journalists, executives—preferred to speak only on background. As indicated in the text, a limited amount of reporting is from trips to Santa Barbara before this project.

## INTRODUCTION

1    **71 percent of young Americans:** Centers for Disease Control and Prevention. (2022, July). *Unfit to serve: Obesity is impacting national security.*

1    **Twelve percent of Americans:** Centers for Disease Control and Prevention. (Last reviewed: 2023, May 15). *Disability impacts all of us.*

1    **About 31 million Americans:** Centers for Disease Control and Prevention. (2018). *Active people, healthy nation: At a glance.*

1    **half of American kids:** Community Preventive Services Task Force. (2020, August 6). *Interventions to increase active travel to school.* The Community Guide.

2    **From 2008 to 2018:** Centers for Disease Control and Prevention. (2021). *Emergency department visits for injury: United States, 2018–2019.* National Center for Health Statistics.

12    **reduce ACL injuries by an incredible 64 percent:** Dos'Santos, T., Thomas, C., and Jones, P. A. (2022). Prevention of non-contact anterior cruciate ligament injuries among youth female athletes: An umbrella review. *International Journal of Environmental Research and Public Health, 19*(7), 4024.

## 1. MEDICINE

17    **Thorndike concocted:** Thorndike, A. (1940). Athletic and related injuries. *New England Journal of Medicine, 223*(5), 160–167.

17    **"a meaty chronicle of sprains":** Medicine: Athletes' injuries. (1940, July 22). *Time.*

18    **lost her life to heart problems:** Hurst, J. W. (1991). Paul Dudley White: The father of American cardiology. *Clinical Cardiology, 14*(7), 622–626.

18    **Harvard Medical School's graduation:** White, P. D. (1957). The evolution of our knowledge about the heart and its diseases since 1628. *Circulation, 15*(6), 915–921.

18    **Robust science began to arrive:** McLean, J. (1925). Official method for lessening heart disease. In *Proceedings of the Annual Meeting of the Association of Life Insurance Medical Directors of America, Volume 10*, 11. Association of Life Insurance Medical Directors of America.

19    **opening a bike path in Chicago:** White, P. D. (1920, September 27). Dr. White opens a Chicago bike path. *Chicago Tribune*, p. 7.

19    **"Death from a heart attack":** Dr. Paul Dudley White is dead at 87; pioneer in care of heart. (1973, November 1). *New York Times.*

19    **"He spoke to the press":** Lee, T. H. (2013). Seizing the teachable moment—Lessons from Eisenhower's heart attack. *New England Journal of Medicine, 368*(23), 2142–2143.

19    **Press secretary Jim Hagerty said:** Lee, T. H. (2013). Seizing the teachable moment—Lessons from Eisenhower's heart attack. *New England Journal of Medicine, 368*(23), 2142–2143.

20    **162 per 100,000 in 2019:** Centers for Disease Control and Prevention. (2021). *Selected mortality tables, 2020* [PDF]. National Center for Health Statistics.

20    **now live, on average, a decade longer:** Centers for Disease Control and Prevention. (2021). *United States life tables, 2019* [PDF]. National Center for Health Statistics.

## 3. WICKED

27  **"His clinical technique included palpating patients' tongues":** Hogarth, R. M., Lejarraga, T., and Soyer, E. (2015). The two settings of kind and wicked learning environments. *Journal of the European Economic Association, 13*(5), 800–829.

28  **legendary sports physiologist Steven Horvath:** Institute plans conference on DNA and aging. (1958, October 3). *Iowa City Press Citizen*, p. 10.

28  **how soldiers acclimate to hot climates:** Robinson, S., Turrell, E. S., Belding, H. S., and Horvath, S. M. (1943). Rapid acclimatization to work in hot climates. *American Journal of Physiology, 140*(2), 168–176.

29  **a leading researcher:** National Institute of Environmental Health Sciences. (1993). Neurodevelopmental effects of PCBs, PCDFs, and PCDDs in humans and animals [PDF]. *Environmental Health Perspectives, 101*(Suppl 2), 103–106.

29  **"ozone had a marked effect on the lung":** Horvath, S. M., and Folinsbee, L. J. (1976). Effects of low levels of ozone and temperature stress. *Environmental Health Perspectives, 13*, 147–152.

33  **People made fun of Michael Jordan:** Hutchinson, A. (2018). *Endure: Mind, body, and the curiously elastic limits of human performance.* William Morrow.

35  **assessed in every imaginable manner:** Gentil, P., Lima, R. M., and Jacó de Oliveira, R. (2013). Superior fatigue resistance of elite black South African distance runners. *International Journal of Sports Physiology and Performance, 8*(5), 458–460.

37  **The ratio came from studies:** Baroni, B. M., Ruas, C. V., Ribeiro-Alvares, J. B., and Pinto, A. S. (2020). Brief review: Hamstring-to-quadriceps torque ratios of professional male soccer players: A systematic review. *Journal of Strength and Conditioning Research, 34*(1), 281–293. Coombs, R., and Garbutt, G. (2002). Developments in the use of the hamstring/quadriceps ratio for the assessment of muscle balance. *Journal of Sports Science & Medicine, 1*(3), 56–62.

37  **"the cause of hamstring injuries is still unclear":** Coombs, R., & Garbutt, G. (2002). Developments in the use of the hamstring/quadriceps ratio for the assessment of muscle balance. *Journal of Sports Science & Medicine, 1*(3), 56–62.

37  **Longitudinal studies:** Ruas, C. V., Pinto, R. S., Haff, G. G., Lima, C. D., Pinto, M. D., and Brown, L. E. (2019). Alternative methods of determining hamstrings-to-quadriceps ratios: A comprehensive review. *Sports Medicine - Open, 5*, 11.

40  **Associated Press estimated:** Walker, T. M., and Fenn, L. (January 28, 2020). AP analysis: NFL teams lost over $500M to injuries in 2019. *AP News*.

40  **"If teams implement reasonably smart hamstring injury prevention":** McMahon, I. (2016, August 17). Why hamstring injuries are so common in NFL players, during preseason training. *SI.com*.

## 4. OLYMPIC MOVEMENT

43  **"it's a lot easier to get to the top":** Partnership a boost for struggling tour. (2003, August 5). *Los Angeles Times*, p. D7.

## 5. THE DUMBEST SPORT

52    **"Nebraska's muscle curtain":** Owens says OU kept in hole by Tom Allan. (1969, October 29). *Omaha World-Herald.*

53    **Epley put on twenty pounds:** Epley, B. (2012). "The Strength of Nebraska": Boyd Epley, Husker Power, and the Making of a Strength Coach. *Journal of Strength and Conditioning Research, 26*(12), 3311–3314. Epley, B. (n.d.). *How Husker Power began with Bob Devaney.* Retrieved from http://www.boydepley.com/1 -TheBeginning.pdf. Power Athlete. (2022, March 4). *Power Athlete Radio Ep 590 // Founder of Husker Power & the NSCA Boyd Epley* [Video file]. Retrieved from https://www.youtube.com/watch?v=H-rL95pcEvg.

53    **"none of whom had necks":** Soucheray, J. (1981, December 13). Weights prove to be uplifting for Huskers. *Lincoln Star*, p. 16.

57    **"the first coach to administer":** USA Strength and Conditioning Hall of Fame. Alvin Roy. Retrieved from https://www.usastrengthcoacheshf.com/member/alvin-roy.

57    **"by this time Riecke realized his spectacular gains":** USA Strength and Conditioning Hall of Fame. Louis Riecke. Retrieved from: https://www .usastrengthcoacheshf.com/member/louis-riecke.

60    **a flood of biomechanical research:** Frost, D. M., Cronin, J. B., Levin, G. (2008). Stepping backward can improve sprint performance over short distances. *Journal of Strength and Conditioning Research, 22*(3), 918–922.

60    **the NSCA published guidelines:** National Strength and Conditioning Association. (2023). *NSCA-CPT study guide.* Human Kinetics.

61    **Ken Griffey Jr. responded:** Baker, G. (2010, February 25). No weighting for Mariners' new conditioning program. *Seattle Times.*

62    **Eric Byrnes told reporters:** Baker, G. (2010, February 25). No weighting for Mariners' new conditioning program. *Seattle Times.*

## 6. LAND

75    **cutting the injury rate by 64 percent:** Mattu, A. T., Ghali, B., Linton, V., Zheng, A., and Pike, I. (2022). Prevention of non-contact anterior cruciate ligament injuries among youth female athletes: An umbrella review. *International Journal of Environmental Research and Public Health, 19*(8), Article 4648.

## 7. JAZZ

77    **fathered a child:** Gee, A. (February 18, 2023). The NBA shouldn't have creepy Karl Malone at All-Star Weekend: His statutory rape of Gloria Bell, absentee fatherhood, and harassment of Vanessa Bryant make him a bad ambassador for the league. *Rolling Stone.*

78    **testing positive for the steroid nandralone:** World Anti-Doping Agency. (2002). *WADA independent observer report on the FIBA World Basketball Championships (Men) 2002.*

## 8. MISOGI

87 **A groundbreaking paper:** Balanoff, A., et al. (2024). Quantitative functional imaging of the pigeon brain: Implications for the evolution of avian powered flight. *Proceedings of the Royal Society B: Biological Sciences, 291*(1825), Article 20232172.

88 **A 2011 study:** Thaler, L., Arnott, S. R., and Goodale, M. A. (2011). Neural correlates of natural human echolocation in early and late blind echolocation experts. *PLoS One, 6*(5), e20162.

88 **Researchers found:** Yong, E. (2011, May 25). The brain on sonar—how blind people find their way around with echoes. *Discover Magazine.*

89 **"exercise, the environment":** Puderbaugh, M., and Emmady, P. D. (2023). Neuroplasticity. In *StatPearls* [internet]. StatPearls Publishing.

89 **"represent the construction":** Peña, J. L., and DeBello, W. M. (2010). Auditory processing, plasticity, and learning in the barn owl. *ILAR Journal, 51*(4), 338–352.

89 **a couple of days:** Roth, E. (n.d.). Neuroplasticity in everyday life. *Brain Waves.* Retrieved from https://biaaz.org/brain-waves/neuroplasticity-in-everyday-life.

89 **"no one laughs at babies":** Dweck, C. S. (2016). *Mindset: The new psychology of success* (Updated paperback ed.). Random House.

90 **"evolutionary, hereditary, developmental":** Toga, A. W., and Thompson, P. M. (2003). Mapping brain asymmetry. *Nature Reviews Neuroscience, 4*(1), 37–48.

91 **"It is hardly necessary to say":** Nishimura, S., and Norinaga, M. (1991). The Way of the Gods: Motoori Noriaga's Naobi no Mitama. *Monumenta Nipponica, 46*(1), 21–41.

93 **"the body's stress response":** Crum, A., and Crum, T. (2015, September 3). Stress can be a good thing if you know how to use it. *Harvard Business Review.*

95 **"The first hour is an eye-stinging, lung-burning":** Bethea, C. (2014, December 9). The one-day-a-year fitness plan: More pain quest than workout, misogi is the secret, punishing ritual that has revolutionized Atlanta Hawks supershooter Kyle Korver's game. You have time for this—if it doesn't kill you first. *Outside Online.*

96 **"We hear a lot about how stress":** Crum, A. J., Salovey, P., and Achor, S. (2013). Rethinking stress: The role of mindsets in determining the stress response. *Journal of Personality and Social Psychology, 104*(4), 716–733.

## 9. MOTION CAPTURE

101 **"We take that poetry":** Recently, P3 used Bayesian machine learning to decipher previously unknown traits that led to knee injury in NBA players. As they prepared the data for publication, they were advised to make a univariate version of the work—which felt a bit like affixing wheels to a hovercraft. The analysis would not offer a better understanding of human bodies, but it would fit more squarely in the tradition of studies.

102 **A 2020 survey of available research:** Lu, L., et al. (2020). Wearable health devices in health care: Narrative systematic review. *JMIR mHealth and uHealth, 8*(11), e18907.

105 **considered recovered from ACL surgery:** Cavanaugh, J. T., and Powers, M. (2017). ACL rehabilitation progression: Where are we now? *Current Reviews in Musculoskeletal Medicine, 10*(3), 289–296.

109 **a classic 149-page study:** Brattström, H. (1964). Shape of the intercondylar groove normally and in recurrent dislocation of patella: A clinical and X-ray anatomical investigation. *Acta Orthopaedica Scandinavica, 35*(sup68), 1–148.

110 **"it is beyond doubt that misalignment":** Khasawneh, R. R., Allouh, M. Z., and Abu-El-Rub, E. (2019). Measurement of the quadriceps (Q) angle with respect to various body parameters in young Arab population. *PLOS One, 14*(6), e0218387.

111 **Other research:** Barber-Westin, S. D., Noyes, F. R., Smith, S. T., and Campbell, T. M. (2009). Reducing the risk of noncontact anterior cruciate ligament injuries in the female athlete. *Physician and Sports Medicine, 37*(3), 49–61. Quatman, C. E., Quatman-Yates, C. C., and Hewett, T. E. (2010). A 'plane' explanation of anterior cruciate ligament injury mechanisms: A systematic review. *Sports Medicine, 40*(11), 729–746. Bisciotti, G. N., et al. (2019). Anterior cruciate ligament injury risk factors in football. *Journal of Sports Medicine and Physical Fitness, 59*(10), 1724–1738. Devana, S. K., Solorzano, C., Nwachukwu, B., and Jones, K. J. (2022). Disparities in ACL reconstruction: The influence of gender and race on incidence, treatment, and outcomes. *Current Reviews in Musculoskeletal Medicine, 15*(1), 1–9.

114 **many different researchers:** Cyron, B. M., Hutton, W. C., Troup, J. D. (1976). Spondylolytic fractures. *Journal of Bone and Joint Surgery British Volume, 58-B*(4), 462–466. Begeman, P. C., Visarius, H., Nolte, L. P., Prasad, P. (1994). Viscoelastic shear responses of the cadaver and hybrid III lumbar response. SAE Technical Paper 942205. Frei, H., Oxland, T. R., Nolte, L. P. (2002). Thoracolumbar spine mechanics contrasted under compression and shear loading. *Journal of Orthopaedic Research, 20*(6), 1333–1338. Bisschop, A., et al. (2012). The impact of bone mineral density and disc degeneration on shear strength and stiffness of the lumbar spine following laminectomy. *European Spine Journal, 21*, 530–536.

114 **"cadaver anterior lumbar failure started":** Gallagher, S., and Marras, W. S. (2012). Tolerance of the lumbar spine to shear: A review and recommended exposure limits. *Clinical Biomechanics, 27*(6), 973–978.

115 **P3 published research:** Rauch, J., Leidersdorf, E., Reeves, T., Borkan, L., Elliott, M., and Ugrinowitsch, C. (2020). Different movement strategies in the countermovement jump amongst a large cohort of NBA players. *International Journal of Environmental Research and Public Health, 17*(Use of Mechanical Variables to Prescribe Training and Evaluate Physical Fitness), Article 6394.

## 10. GRAVITY

117 **a 1988 *New Yorker* story:** McPhee, J. (1988, September 26). Los Angeles against the mountains. *The New Yorker.*

121 **LaVine needed his mom to calm him down:** Smith, S. (2020, January 30). Now is the time for Zach LaVine to be an All-Star. *NBA.com.*

123　**one of the biggest wildfires in California history:** Ventura County Fire Department. (2020, March 13). VNCFD determines cause of the Thomas Fire.

126　**Zoffness tells a story:** Fisher, J. P., Hassan, D. T., O'Connor, N. (1995). Minerva. *BMJ, 310*(70).

128　**"Results indicate that the 'game' ":** Hastorf, A. H., and Cantril, H. (1954). They saw a game; a case study. *Journal of Abnormal and Social Psychology, 49*(1), 129–134.

## 11. GROUND UP

131　**"The colossal load that acts":** Verkhoshansky, Y. (2018). *Shock Method*. Retrieved from https://www.verkhoshansky.com/Portals/0/Book/SM_Index.pdf.

## 12. TEN MILLIMETERS FROM NOWHERE

147　**Demesse Tefera:** Sometimes spelled Demese (see, for example, https://worldathletics .org/athletes/ethiopia/demesse-girma-14181653), but his home club, the West Side Runners, spells it "Demesse."

148　**"can definitely make things worse":** Carroll W. Basketball shoe trends favor fashion over feet. (2017, February). *Lower Extremity Review.*

148　**a 2022 survey of running shoe studies:** Relph, N., et al. (2022). Running shoes for preventing lower limb running injuries in adults. *Cochrane Database of Systematic Reviews, 8*(8).

148　**"the analysis found no evidence":** Kuzma C. (2023, March 30). What you do (and don't) need in a running shoe. *New York Times.*

150　**pronghorn antelope:** Fischer, H. (2008, June 20). The pronghorn's prowess. *Discover Magazine.*

152　**"The model when building running shoes":** McMahon, I. (2016, June 8). Ampla Fly shoe: A rocket with carbon fiber wings. *Sports Illustrated.*

## 13. DRAGON'S BACK

162　**plausible in the research:** Fox, F. A. U., Liu, D., Breteler, M. M. B., and Aziz, N. A. (2023). Physical activity is associated with slower epigenetic ageing— Findings from the Rhineland study. *Aging Cell, 22*(6), e13828.

162　**"Pain is not just bad in itself":** Setiya K. (2022, November 1). Why does chronic pain hurt so much? *Atlantic.*

163　**we are evolved from fish:** Huberman, A. (Host). (2022, June 19). Ido Portal: The science & practice of movement [Audio podcast episode]. Retrieved from https:// www.youtube.com/watch?v=a9yFKPmPZ90&t=1s

165　**don't cause pain:** Brinjikji, W., et al. (2015). Systematic literature review of imaging features of spinal degeneration in asymptomatic populations. *American Journal of Neuroradiology, 36*(4), 811–816.

## 14. GET INTO YOUR HIPS

166　**looked like "crab meat":** Torre, P. S. (2016, February 29). How Stephen Curry got the best worst ankles in sports. *ESPN The Magazine.*

168　**In his book:** DeSilva, J. (2022). *First steps: How upright walking made us human.* Harper.

170　**TFL has an incredible task list:** Trammell, A. P., Nahian, A., and Pilson, H. (2023). Anatomy, bony pelvis and lower limb: Tensor fasciae latae muscle. *Stat-Pearls* [Internet]. Last updated August 17, 2023.

170　**"It is difficult":** Bordoni, B., and Varacallo, M. (2023). Anatomy, abdomen and pelvis, quadratus lumborum. *StatPearls* [Internet]. Last updated July 17, 2023.

171　**"30 fresh-frozen cadaveric knees":** Ueno, R., et al. (2020). Analysis of internal knee forces allows for the prediction of rupture events in a clinically relevant model of anterior cruciate ligament injuries. *Orthopedic Journal of Sports Medicine,, 8*(1).

177　**"basketball players spend approximately 31 percent":** Leidersdorf, E., et al. (2022). Reliability and effectiveness of a lateral countermovement jump for stratifying shuffling performance amongst elite basketball players. *Sports, 10*(11), 186.

178　**most ferocious growth areas:** Shichman, I., et al. (2023). Projections and epidemiology of primary hip and knee arthroplasty in Medicare patients to 2040–2060. *Journal of Bone and Joint Surgery Open Access, 8*(1), e22.00112.

178　**A 2023 *Journal of Rheumatology* article:** Singh, J. A., Yu, S., Chen, L., and Cleveland, J. D. (2019). Rates of total joint replacement in the United States: Future projections to 2020–2040 using the National Inpatient Sample. *Journal of Rheumatology.*

178　**$40,000 each:** Paretts, S. (2023, August 25). What is the average cost of hip replacement surgery? *Well U.* Retrieved from https://www.carecredit.com/well-u/health-wellness/hip-replacement-surgery-cost/.

## 15. PSOAS

182　**Studies have correlated:** Fitzpatrick, J. A., et al. (2020). Large-scale analysis of iliopsoas muscle volumes in the UK Biobank. *Scientific Reports, 10,* 20215.

182　**lumbar plexus:** Ocran, E., and Grujičić, R. Lumbar plexus. In: Kenhub. Last reviewed: October 30, 2023.

182　**highly accurate ways:** Holzbaur K. R. S., Murray, W. M., Gold, G. E., and Delp, S. L. (2007). Upper limb muscle volumes in adult subjects. *Journal of Biomechanics, 40,* 742–749.

183　**407 cubic centimeters:** Fitzpatrick, J. A., et al. (2020). Large-scale analysis of iliopsoas muscle volumes in the UK Biobank. *Scientific Reports, 10,* 20215.

183　**"An estimated 619 million people":** World Health Organization. (2023, June 19). *Low back pain.* Fact Sheet [Internet]. Geneva: World Health Organization.

183　**A growing body of research:** Pourahmadi, M., Asadi, M., and Yeganeh, A. (2020). Changes in the macroscopic morphology of hip muscles in low back pain. *Journal of Anatomy, 236*(1), 3–20. Lube, J., et al. (2016). Reference data on muscle volumes of healthy human pelvis and lower extremity muscles: an in vivo magnetic resonance imaging feasibility study. *Surgical and Radiologic Anatomy, 38*(1), 97–106. Frącz, W., Matuska, J., and Skorupska, E. (2023). The cross-sectional area assessment of pelvic muscles using MRI manual segmentation among patients with low back pain and healthy subjects. *Journal of Imaging, 9*(8), Article 155.

184 **increases in psoas size:** Yamanaka, R., Wakasawa, S., Yamashiro, K., Kodama, N., and Sato, D. (2021). Effect of resistance training of psoas major in combination with regular running training on performance in long-distance runners. *International Journal of Sports Physiology and Performance, 16*(6), 906–909.

186 **"LIFE CHANGING":** SpineCare Decompression and Chiropractic Center. (2022, March 28). *How to fix a tight psoas muscle in 30 SECONDS* [Video file]. Retrieved from https://www.youtube.com/watch?v=kibVUeXFmwA.

### 16. RELAX

191 **"Fast and relaxed is the name of the game.":** Magness, S. (2023, May 23). Sprinting teaches us a great lesson that can applied to almost all of life. [Twitter post]. Retrieved from https://twitter.com/stevemagness/.

192 **thirty to ninety seconds:** Fletcher, J. (2023, November 8). How long can the average person hold their breath? *Medical News Today.*

192 **neural response to elevated CO2 is poorly understood:** Shigemura, M., and Sznajder, J. I. (2021). Elevated CO2 modulates airway contractility. *Interface Focus, 11*(2), 20200021.

### 17. RATTLERS

206 **$100 billion a year on sports:** Greenberg, D. (2024, February 20). Sports betting industry posts record $11B in 2023 revenue. *ESPN.*

206 **$200 million a quarter into start-ups:** Glasner, J. (2024, April 22). Investors re-engage with gaming startups. *Crunchbase News.*

209 **741 injuries a year:** Drakos, M. C., Domb, B., Starkey, C., Callahan, L., Allen, A. A. Injury in the National Basketball Association: A 17-year overview. *Sports Health, 2*(4), 284–290. "In the current study, 12 594 injuries in 1366 players occurred over 17 years." Dividing 12,594 injuries by seventeen years yields an average of 741.

### 18. SLOW

220 **a 1968 World Health Organization paper:** Wilson, J. M. G., and Jungner, G. (1968). *Principles and practice of screening for disease.* (Public Health Papers, No. 34). World Health Organization.

### 19. GREAT HORNED OWL

224 **50 million people read his story:** Tan, S. Y., and Kwock, E. (2016). Paul Dudley White (1886–1973): Pioneer in modern cardiology. *Singapore Medical Journal, 57*(4), 215–216.

224 **"I myself," writes White:** White, P. D. (1955, October 30). Heart ills and presidency: Dr. White's views. *New York Times.*

224 **"it is time for us at all ages":** White, P. D. (1957, June 23). Rx for Health: Exercise. *New York Times.*

225 **President's Council on Youth Fitness:** Health.gov. (n.d.). History of the President's Council on Physical Fitness and Sports. Retrieved June 14, 2024.

225 **American Heart Association recommended:** Troiano, R. P. (2016, July 14). *History of physical activity recommendations and guidelines for Americans.* National Cancer Institute, National Institutes of Health.

225 **"vigorous-intensity aerobic physical activity":** Physical Activity Guidelines Advisory Committee. (2024). *Physical activity guidelines for Americans.* health.gov.

225 **less than a quarter of Americans:** Elgaddal, N., Kramarow, E. A., and Reuben, C. (2022). *Physical activity among adults aged 18 and over: United States, 2020* (NCHS Data Brief No. 443). National Center for Health Statistics.

226 **a newspaper story:** McCarthy, A. (2023, March 25). Whatever the problem, it's probably solved by walking. *New York Times.*

230 **An Oregon nature writer:** Anderson, J. (2018, December 19). A brush with a papa owl. *Bend Source.*

## 20. STORK PRESS

235 **hard for Kerr to stand:** Shelburne, R. (2016, April 7). Steve Kerr has suffered more than you will ever know. *ESPN.*

236 **"associated with significant increase":** Nguyen, T. H., Randolph, D. C., Talmage, J., Succop, P., and Travis, R. (2011). Long-term outcomes of lumbar fusion among workers' compensation subjects: A historical cohort study. *Spine, 36*(4), 320–331.

236 **"even in young adults, degenerative changes":** Brinjikji et al. (2015).

## 21. SEVEN FALLS

249 **A study in India:** Dhawale, N., and Venkadesan, M. (2023). How human runners regulate footsteps on uneven terrain. *eLife, 12*, e67177.

# Index

Page numbers in *italics* refer to illustrations.
ME = Marcus Elliott; HA = Henry Abbott.